MOLECULAR BIOLOGY OF ONCOGENES AND CELL CONTROL MECHANISMS

ELLIS HORWOOD SERIES IN MOLECULAR BIOLOGY

Series Editor: Dr A. J. TURNER, Professor of Biochemistry, University of Leeds

MOLECULAR BIOLOGY OF ONCOGENES AND CELL CONTROL MECHANISMS

Editors

PETER J. PARKER B.A., Ph.D.
Principal Scientist, Imperial Cancer Research Fund, London

MATILDA KATAN B.Sc., M.Sc., Ph.D.
Research Scientist, Marie Curie Research Institute, Oxted

ELLIS HORWOOD
NEW YORK   LONDON   TORONTO   SYDNEY   TOKYO   SINGAPORE

First published in 1990 by
ELLIS HORWOOD LIMITED
Market Cross House, Cooper Street,
Chichester, West Sussex, PO19 1EB, England

A division of
Simon & Schuster International Group
A Paramount Communications Company

Printed and bound in Great Britain
by Bookcraft (Bath) Limited, Midsomer Norton, Avon

British Library Cataloguing in Publication Data

Molecular biology of oncogenes and cell control mechanisms
(Ellis Horwood series in molecular biology)
Parker, P. J. and Katan, M.
CIP catalogue record for this book is available from
the British Library
ISBN 0–13–599499–3

Library of Congress Cataloging-in-Publication Data

Molecular biology of oncogenes and cell control mechanisms /
editors, Peter J. Parker, Matilda Katan
p. cm. — (Ellis Horwood series in molecular biology)
ISBN 0–13–599499–3
1. Cellular control mechanisms. 2. Oncogenes. 3. Growth
factors. 4. Second messengers (Biochemistry) 5. Cell
receptors. 6. Molecular biology. I. Parker, Peter J., 1954– .
II. Katan, Matilda, 1954– . III. Series.
QH604.M64   1990
616.99′4071–dc20   90–44157
                                                          CIP

Table of contents

Preface

The amazing progress and exciting achievements which have taken place in recent years in our understanding of the signal transduction pathways initiated by hormones and growth regulators has revolutionized our understanding of the control, normal development and origins of many human diseases - particularly cancer. These advances have largely been made possible by innovative experimental approaches in molecular cell biology which exploit genetic engineering. This book succeeds in capturing our current knowledge of signal transduction processes and it does so because it is written by those actively involved in deciphering the molecular anatomy of the interacting component parts. By identifying the key areas which have led to our current understanding and describing the elements which remain to be characterized, a remarkable insight into the future of a very fast developing field is presented.

There are many reasons why we have seen an explosive growth in our knowledge of the signal transduction processes. Not the least of these is the interaction which has occurred between researchers in the different sub-specialties which are now grouped in molecular cell biology. The ability to purify extremely small amounts of the enzymes and proteins involved in the signal transduction process, particularly of course the initiator growth regulators and their receptors, together with the advances in analytical biochemistry which can provide sequence information for the cloning of those molecules, and perhaps even more importantly, the ability to derive clones through an analysis of the expressed function of those clones, has led to the generation of a vast amount of primary sequence information both at the protein and DNA level. The interpretation of this data depends of course on information about the function of structural motifs often identified initially by computer analysis of the ever growing data banks. The ability to introduce into cells the genes which encode identified components, or to knock out these components, will make possible further and remarkable progress in sorting out the key steps in the signal transduction process. It is the field of cancer research which has provided some of the key discoveries which have advanced our knowledge of signal transduction and allowed us to identify key steps which are subject to oncogenic subversion. The elucidation of the function of certain oncogenes as key players in the signal transduction process and the identification of tumour suppressor genes have been some of the most exciting advances in cancer research in the last few years. The authors have played key roles in some of these discoveries which are covered in detail in this book.

The rewards for new discoveries that remain to be made in this area of research are conveyed in the exciting, informative text of the book. I have no doubt this will serve to stimulate new researchers to fearlessly try and fill the gaps in our knowledge.

M D Waterfield
Professor of Experimental Oncology and Director of Research, Ludwig Institute for Cancer Research, University College and Middlesex School of Medicine.

Abbreviations; some common abbreviations

aFGF	acidic fibroblast growth factor
AMD phosphatase	ATP-Mg-dependent phosphatase
AMD_c, F_c	AMD catalytic subunit
ANP	atrial natriuretic peptide
AR	adrenergic receptor
ATF	activator transcription factor
βARK	β-adrenergic receptor kinase
bFGF	basic fibroblast growth factor
bZip	basic leucine zipper
cAkinase (cAPK)	cAMP-dependent protein kinase
CAM	calmodulin
CAMK	Ca^{2+}-calmodulin-dependent protein kinase
cAMP	cyclic AMP
cDNA	complementary deoxyribonucleic acid
cGMP	cyclic GMP
cGPK	cGMP-dependent protein kinase
CK	Casein kinase
CRE	cAMP response element
CREB	CRE binding protein
CSF	colony stimulating factor
DG,DAG	diacylglycerol
DLAR	*drosophila* LAR
DNA	deoxyribonucleic acid
DPTPase	*drosophila* PTPase
EGF	epidermal growth factor
EGFR	EGF receptor
GAP	GTPase activating protein
G.C.A.T.	guanosine. cytosine. adenosine. thymidine
GHRF	growth hormone releasing factor
G protein	GTP binding protein
GSK3, F_A	glycogen synthase kinase3
HD	homeodomain .
HER2	human EGF receptor 2
HSE	heat shock element
IGF	insulin-like growth factor
IP_3	inositol 1,4,5 trisphosphate
LAR	leukocyte common antigen related protein

M	modulator subunit
MLCK	mysoin light chain kinase
NGF	nerve growth factor
PCS phosphatase	polycation stimulated phosphatase
PDGF	platelet derived growth factor
PhK	phosphorylase kinase
PI-PLC	inositol lipid-specific phospholipase C
PIP$_2$	phosphatidyl inositol 4,5 bisphosphate
PKC	protein kinase C
PR subunits	phosphatase regulatory subunits
PTPase	phosphotyrosine protein phosphatase
RB	retinoblastoma
RNA	ribonucleic acid
RPTPase	receptor-like PTPase
SH2/3	*src* homology regions 2/3
TGF	transforming growth factor
TPA	tetradecanoyl phorbol acetate
TRE	TPA response element
VVGF	vaccina virus growth factor

1

Introduction

1. A general framework of signal transduction

The term signal transduction is in essence a replacement for its predecessor, the proverbial "black box". It has been known for many years that hormones and factors bind to receptors on the surface of cells and that as a consequence "something happens" leading to a specific cellular response. Within the last 20 years significant progress has been made in trying to open this box and as a result, many different signalling systems and regulatory pathways are now emerging from the darkness.

Figure 1 outlines the essential steps involved in regulation of glycogen breakdown by β-adrenergic agonists. This particular sequence of events has provided a conceptual corner stone in the elucidation of other related and unrelated transduction systems. Transduction from the β-adrenergic receptor to phosphorylase b as indicated in figure 1, illustrates various specific steps in transduction processes that can be formulated into a generalised transduction scheme (figure 2). This generalised signal transduction pathway provides a framework in which a number of distinct pathways can be discussed. Following the general transduction sequence (figure 2), the first step is recognition of an extracellular signal. The extracellular signals are generated by a range of changes in cellular environment including changes in the concentration of hormones or alterations to extracellular matrix or to the surfaces of neighbouring cells. In order to recognize these extracellular signals, cells express receptors. The presence or absence of a particular receptor determines whether a signal is perceived and as a consequence whether a response is set in motion. Some agents, such as retinoids and steroids, can penetrate the plasma membrane and bind to their receptors located in the cytoplasm or nucleus. However, the majority of signals are not membrane permeant and their receptors are present at the cell surface.

Interaction between cell surface receptors and specific signals leads to a change in receptor function loosely described as "receptor activation". An "active" receptor is capable of generating an intracellular signal that is further transmitted through the cell. The transmission of the receptor generated signal often employs production of small diffusable molecules, called second messengers, and transient covalent modifications of proteins, in particular protein phosphorylation. Both the binding of a second messenger and attachment of a phosphate group to a protein component of

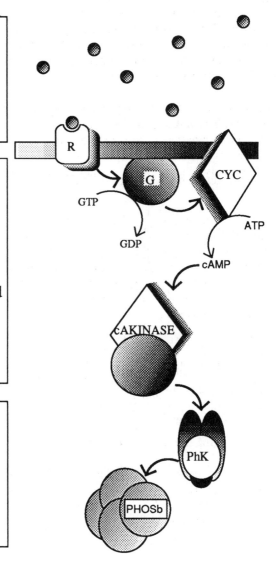

RECOGNITION OF
AN EXTERNAL
SIGNAL (e.g.
hormone receptor
interaction)

TRANSDUCTION
OF THE SIGNAL
ACROSS THE
MEMBRANE AND
THROUGH THE
CYTOPLASM (e.g.
production of second
messenger and
changes in protein
phosphorylation)

CHANGE IN
FUNCTION OF
A PROTEIN (e.g.
enzymes,
transcription
factors)

Fig 1. A paradigm for signal transduction. The processes illustrated here are those involved in the hormonal regulation of glycogenolysis. This classic example of intracellular signal transduction involves the following chain of regulatory components: hormone, adrenaline; transmembrane receptor, β-adrenergic receptor (R); coupling G-protein, (G); effector enzyme, adenylyl cyclase (CYC); second messenger, cyclic adenosine monophosphate (cAMP); intracellular second messenger receptor, cAMP-dependent protein kinase (cAKINASE); primary target, phosphorylase kinase (PhK); secondary target, phosphorylase b (PHOSb).

signal transduction can modify its specific function. To date, only a few second messengers have been discovered, however, these second messengers are a common component of many different regulatory pathways. In contrast, the number of different protein kinases and phosphatases, enzymes that catalyse protein phosphorylation and dephosphorylation, is relatively large. This is particularly true for the protein kinases, which can participate in signal transduction as targets for second messengers and also at other levels of signalling hierarchies.

The final step in this chain of signalling reactions is an alteration in the behaviour of target proteins. The target proteins vary from rate limiting enzymes of relatively "simple" metabolic pathways to the factors that regulate expression of one or more genes, critical for cell proliferation or differentiation. As a consequence the cellular 'response' can vary from a switch in fuel metabolism to a gross alteration in cell function (eg. differentiation of a stem cell).

During transduction of an extracellular signal there are steps that are responsible for amplification of the signal. This amplification occurs when one component stimulates, or produces, many molecules of another signalling component, i.e. where a transducer component acts catalytically. This is of course an extremely important element in any signalling pathway where the concentration of, for example, a hormone is orders of magnitude lower than the concentration of the ultimate target proteins that it impinges upon.

In addition to the chain of reactions from an external signal to target proteins (feed forward), many pathways include inhibitory, feed back loop(s). Furthermore there are "cross-talk sites" at which products of other regulatory pathways can modulate the flow of information. Signal transduction pathways can interact at multiple levels, with both positive (cooperation) and negative (inhibition) consequences. Thus, a complex regulatory network in the cell transmits, amplifies and coordinates an array of signals impinging upon the cell surface.

Within the general framework of signal transduction described above, recognition of a signal, its transduction to a target protein and integration into a cellular regulatory network can be achieved in many different ways. The complexity of the pathways can vary from those that employ only a few components, to pathways that consist of a long chain of participants (figure 2).

2. A unit of signal transduction: the generic transducer protein

Protein components that participate in signal transduction, such as receptors, protein kinases or transcription factors, transduce many different signals and come in a variety of shapes and sizes. However, they all perform the same basic functions (figure 3) and these proteins can be collectively called generic transducers.

A generic transducer detects a specific input signal (detector function) and responds to the input signal by generating a specific output signal (generator function). The

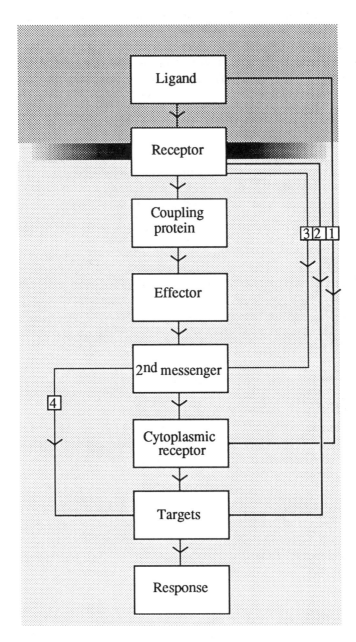

Fig 2. A generalised signal transduction pathway. The figure illustrates a simplified linear pathway from the recognition of an extracellular signal (ligand-receptor interaction) to an intracellular response. The steps illustrated essentially follow those employed in the β-adrenergic regulation of phosphorylase b (figure 1). However also indicated are alternative strategies that can be employed following other types of ligand-receptor interactions; these by-pass certain steps. For example (i) steroid hormones are membrane permeant and interact directly with intracellular receptors [1]; (ii) there is evidence that certain growth factor receptors can directly regulate intracellular proteins [2]; (iii) some receptors have intrinsic "effector" capacity i.e. directly produce second messengers [3]; (iv) certain second messengers act pleiotropically and interact with a number of target proteins to produce a coherent integrated response [4].

output signals can be detected by another transducer as an input signal and thus provide a link between two units. In addition to the obligatory detector and generator function, some transducers also have modulator and/or timer functions. The modulator function allows the transducer to detect and respond to a signal distinct from the primary input signal. The fourth function, the timer, regulates the duration of the response to the input signals.

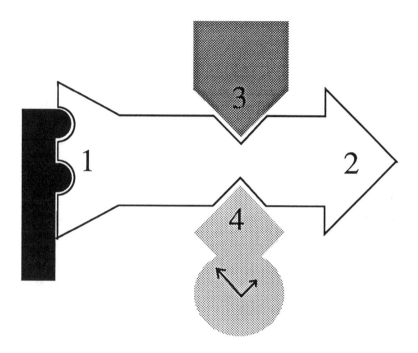

Fig 3. A generic tranducer. The figure illustrates the four functions associated with transducer proteins. These functions are 1. - a detector that perceives an upstream event, 2. - a generator that provides the flow from upstream to downstream events, 3. - a modulator that is responsible for regulating the effectiveness of the transducer and 4. - a timer which provides a self-limiting element to the transducer. Not all transducers have a modulator or timer function. (According to *Bourne/ 1988/ CSH Symp. Quant. Biol. vol. LIII*).

Several of the transducers illustrated in figure 1 provide good examples of their common functional properties. Receptor, G-protein, adenylyl cyclase, cAMP-dependent protein kinase, phosphorylase kinase and the target enzyme phosphorylase are connected into a single pathway by input and output signals. The detector part of the receptor binds hormone and this binding provides an input signal. A similar example is afforded by the binding of cAMP to the cAMP-dependent protein kinase. However, input signals are not restricted to the binding of a ligand to a detector element of a transducer. Protein-protein interactions or protein phosphorylations can also generate an input signal. Examples of protein-protein interactions are afforded by the interaction of receptor with G-protein and also of G-protein with adenylyl cyclase. Protein phosphorylation as an input signal is illustrated by the phosphorylation of phosphorylase catalysed by phosphorylase kinase and of phosphorylase kinase itself.

All these input signals induce conformational changes in the transducer that causes a change in the rate at which its generator part produces a characteristic output signal. For example, an input signal detected by adenylyl cyclase changes the rate of cAMP synthesis, while in the case of the cAMP-dependent protein kinase it changes the rate of substrate phosphorylation. Output signals (like input signals) can include association with or dissociation from another protein.

To illustrate another function of generic transducers, modulator function, a good example is provided by the β-adrenergic receptor. In addition to the binding of hormone, this receptor responds to secondary inputs, including phosphorylation by a specific protein kinase (β- adrenergic receptor kinase). The phosphorylation of the β-adrenergic receptor by this kinase dampens the signal transduction by making the receptor unavailable for the continuing presence of the primary input, i.e. hormone. In general, transducers that respond to more than one signal provide cross-talk sites between different signalling pathways.

The timer function of a generic transducer is probably best understood in the case of G-proteins. The conformation of the G-protein that can generate an output signal requires GTP to be bound to the protein. The timer part of this transducer is its intrinsic GTP-ase activity. Hydrolysis of GTP into GDP (a zero-order event) takes tens of seconds and limits production of the output signal to this period of time. In order for the G-protein to participate further in signal transduction it must again interact with receptor and exchange GTP for GDP.

3. Control of cellular growth, oncogenes and anti-oncogenes

A major challenge in the field of signal transduction is to understand the complex signalling machinery that allows a cell to detect extracellular changes in order to make relevant decisions about proliferation/quiescence and differentiation. It is not, at present, possible to rationalize this signalling system as a complete circuit diagram of interconnected pathways. Many links are still missing and few pathways can be traced through an unbroken chain of identified components all the way from an extracellular signal to for example, a factor that regulates expression of a specific gene(s) that is itself important in regulating cell division.

Notwithstanding the above, a number of signalling pathways have been identified. Their roles in the control of proliferation and differentiation are complex since it appears to be important in what context a pathway is activated (i.e. cell type) and also what other parallel pathways are also stimulated. Thus any one pathway may not be obligatory although cooperation between pathways is often required for a maximum response. This flexibility in selecting a pathway, or rather, a combination of pathways underlies, at least in part, the elaborate and complicated nature of the regulatory network.

In spite of our poor understanding of the complexity of mitogenic signalling, a large number of the transducer proteins are already known. Much information about these transducer proteins has come from studies of their aberrant versions, proteins encoded by oncogenes. So far, about 50 oncogenes have been isolated either from tumour cells or from recombinant transforming retroviruses (figure 4). Oncogenes have been identified following their ability to change normal cells into transformed

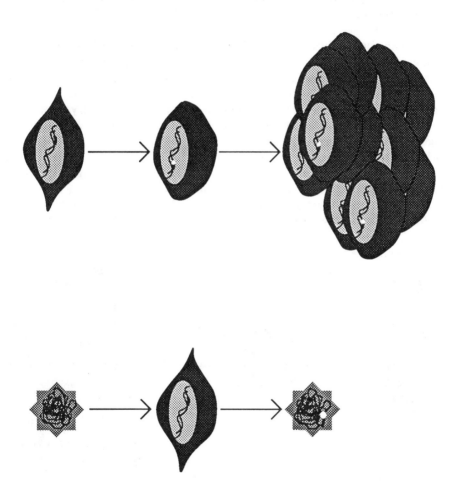

Fig 4. Generation of oncogenes. In the upper section of the figure, a normal gene (heavy line) is shown to undergo a mutation (white line) which leads to a derestriction in cell growth. Such mutations can be induced by carcinogens. In the lower section, a retrovirus is shown infecting a normal cell and as a consequence incorporating and mutating a cellular gene. Following infection by this altered virus, the viral genetic information including the oncogene can become integrated into the host DNA, leading to derestricted growth.

cells *in vitro*, and cause tumours *in vivo*. Analysis of these oncogenes has demonstrated that they originate from cellular genes, the proto-oncogenes, that encode components important in normal growth control. This identification with a particular type of signalling molecule has been used as a basis for classification of oncogenes. Thus, there are oncogenes related to: (i) growth factors, (ii) receptors, (iii) GTP-binding proteins, (iv) protein kinases, and (v) nuclear proteins involved in the regulation of gene expression.

There are two major routes by which a proto-oncogene is converted into an oncogene. The first is deregulation of gene expression; changes that occur affect the level of the gene product. An increase in the number of molecules of a signal transducer can promote extensive or inappropriate proliferation. The second general means of proto-oncogene activation involves changes that affect the structure (and consequently function) of the encoded protein. These structural changes, in one way or another, cause a transducer protein to become "locked" in a conformation that uncontrollably produces an output signal.

In addition to the mechanisms that promote growth, the regulatory network also includes mechanisms that suppress or constrain cell growth. The influence of signalling molecules involved in the latter processes becomes noticeable when they are lost or inactivated. Genes that encode these signalling molecules have been designated as "growth suppressor genes" or "anti-oncogenes" (figure 5). Known anti-oncogenes encode nuclear proteins involved in the regulation of gene expression. Other types of signal transducers (for example the protein phosphatases) may also have a role as growth suppressors.

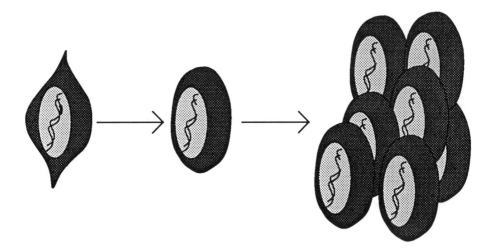

Fig 5. Loss of anti-oncogenes causes unrestricted proliferation. Certain genes appear to be important in positively restricting cell proliferation. The consequence of deletion of such a gene (heavy line) is unrestricted growth as illustrated here. Such loss of function usually is recessive and both alleles have to be affected, however it is possible that dominant mutations could also occur in anti-oncogenes.

Although creation of a single oncogene or loss of an anti-oncogene appears to be critical for genesis of a tumour, it is not necessarily sufficient. Tumourigenesis is a much more complex multistep phenomenon and the processes involved are beyond the scope of the molecular mechanisms described in this book. Here we will describe different types of signal transducers, namely, receptors, second messenger systems (GTP-binding proteins and second messenger producing enzymes), protein kinases, protein phosphatases, and nuclear proteins involved in the regulation of gene expression. Particular emphasis is placed on those transducers that are involved in the regulation of cellular growth, since it is these elements that appear to be altered in malignancy.

2

Polypeptide growth factors

1. Growth factors

Growth factors can be defined as agents that stimulate cells to divide. In general, growth factors show potent biological activities, and are typically active at pM to nM concentrations.This contrasts with nutritional requirements for growth and division which are frequently required at much higher concentrations. In some cases it can be difficult to distinguish between growth factor activity and nutritional requirements. An example is the case of the iron binding protein transferrin, which when added to cells limited for growth by their supply of iron may appear to stimulate growth as a *bona fide* growth factor.

Although we will deal here with polypeptide growth factors, there are other molecules which have been found to stimulate cell division under certain circumstances, including, steroids, cAMP analogues, lectins, and more recently neurotransmitters. In general, the range of structures known to stimulate mitogenesis is broadening, as previously unsuspected growth stimulatory activities of non-polypeptide factors are recognised.

1.1 The discovery of polypeptide growth factors

Modern growth factor research began with the discovery that mouse submaxillary glands contain large quantities of a substance which when injected into newborn mice causes specific developmental changes. These changes seemed to occur via the enhancement of epidermal growth, and included early eruption of the incisors and precocious opening of the eyelids. After purification this factor was found to also stimulate the growth of epidermal and epithelial cells in culture, and was named epidermal growth factor (EGF).

The discovery of another well known growth factor was linked to similar growth stimulatory (mitogenic) effects on cells in culture. Although fibroblasts are able to survive in culture dishes in synthetic media containing plasma, they do not grow well unless serum is added (figure 1). Plasma is obtained after removing the cellular constituents from blood, while serum is derived from whole blood in which platelets have released many wound repair factors and aggregated to form a clot. It was therefore suggested that blood platelets contain a factor which could stimulate the growth of fibroblasts. Ten years later such a factor was indeed purified from human platelets and named platelet-derived growth factor (PDGF).

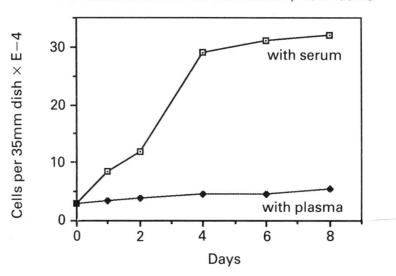

Fig 1. The dependence of fibroblasts on serum. The growth curves illustrate the principles of using a cell proliferation assay to measure growth factor activity. Chicken fibroblasts were grown in synthetic medium containing either chicken plasma or serum prepared from the same sample of blood. The amount of growth was measured after the elapsed time indicated by counting the numbers of cell present. [Taken from *S.D.Balk, Proc.Natl.Acad.Sci.U.S.A. 68, 271-275 (1971)*]

Many cultured cells require serum for growth, in some instances because it may simply provide molecules required for nutrition but missing from synthetic growth media, such as for example the transport protein transferrin which plays an important role in iron uptake as mentioned above. However in many cases the serum provides specific growth factors derived from platelets, including platelet-derived endothelial cell growth factor, an epidermal growth factor-like protein, insulin-like growth factors I and II, and transforming growth factor β, in addition to PDGF.

Numerous growth factors have now been identified and purified from a wide range of sources, including adult tissues, foetal tissues, and media conditioned from cultured cells (including continuous and tumour-derived lines).

The isolation of factors from tumour cell lines raised the important question of whether the abnormal cells were making new factors not found in normal cells, or whether such factors were identical or related to known growth factors. Although detailed structural analyses are not available for all tumour-derived factors, it is clear that most tumour growth factors are closely related, although not necessarily identical, to growth factors isolated from normal tissues.

Two of the best understood cases, involving discoveries linking oncogenes implicated in cancer with growth factors are discussed below. These discoveries together with earlier studies relating the abnormal growth of tumour cells to the abnormal production of growth factors have strongly supported an important working model of how tumour growth factors may function: the autocrine hypothesis. This hypothesis proposes that the abnormal production and secretion of a growth factor by a cell which has receptors for that factor, transforms such a cell into one capable of independent replication, that is, a tumour cell (figure 2). Plausible variations on this theory include paracrine and endocrine effects, and the possibility that the growth

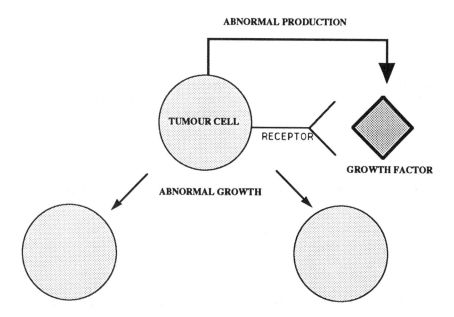

Fig 2. The autocrine hypothesis. The autocrine hypothesis proposes that the abnormal growth of some tumour cells is a consequence of the abnormal secretion of a growth factor, which can then bind to specific receptors on the cell surface. Such a cell thus receives a continuous proliferation signal, and may then be independent of normal growth control mechanisms.
Variations on this model are also plausible, such as paracrine stimulation, where the secreted growth factor may act on adjacent cells other than the secreting cell, and endocrine stimulation, where the secreted growth factor may be carried via the blood stream to distant sites to affect other cells. Another proposed variation of the autocrine hypothesis suggests that the abnormally produced growth factor may not need to be secreted from the cell, but might interact with its specific receptor intracellularly, to generate a proliferation signal.

factor may not necessarily need to be secreted. It is important to note however that human cancer is thought to develop from multiple abnormalities rather than a single change and autocrine growth may well contribute only in part to certain cancers.

There are now more than thirty well characterized and distinct polypeptide growth factors known and probably many more new factors as yet only partially purified or undiscovered. Fortunately, from the properties of the members of the presently well studied factors, there are a number of clear general principles which emerge.

2. General characteristics of growth factors

2.1 Purification

Most growth factors are in general present in very low concentrations in tissues or conditioned media, typically in ng/g tissue (or ml medium) quantities with some notable exceptions, such as EGF in mouse submaxillary gland which can occur at mg/g levels. It is almost impossible to overemphasize the importance of purifying factors to homogeneity, to ensure that the properties of the factor being studied are not due to contaminating proteins. However it is equally hard to overemphasize the difficulties experienced in purifying some growth factors, which as indicated above, are often present in one part in a million or less in the starting material. Large quantities of difficult to obtain tissue or conditioned medium may be needed to isolate a sufficient quantity of the pure protein for study (Table 1). There are many well known growth factors whose first purification to the point of obtaining amino acid sequence has taken more than ten years work by many people from several laboratories. Two important factors, transforming growth factor α and transforming growth factor β, were originally isolated from scarce sources and studied as an apparently pure "transforming growth factor" preparation, before it was recognized that such preparations in fact contained mixtures of these two very different molecules.

The difficulties in obtaining sufficient quantities of material for structural and bio-logical studies are increasingly being overcome through the production of growth factors using recombinant DNA technology. Although the production, purification and characterization of recombinant factors is not trivial, requiring considerable information best established using the naturally occurring proteins, this approach may be essential for many factors normally expressed at extremely low levels.

One of the most frustrating problems in following growth factor research is in understanding the confusing and often inconsistent nomenclature. This should not be taken as a limitation of the field, but rather as a reflection of the tremendously complex tasks of purification and identification of some of these new proteins. Such problems are compounded by the difficulty of establishing the often multiple biological properties of the factors. Alternative names have been included here wherever possible.

FACTOR	SOURCE	STARTING MATERIAL	YEAR	YIELD µg
PDGF	Human platelet-rich plasma	250 L blood	1979	18
PDGF	Fresh human platelets	600 L blood	1981	500
PDGF	Human platelet-rich plasma	200 L blood	1981	760
PDGF	Human platelet-rich plasma	40 L blood	1982	36
PDGF	Fresh porcine platelets	1000 L blood	1984	266
TGF-β	Bovine kidney	14 KG	1983	40
TGF-β	Human placenta	8.8 KG	1983	88
TGF-β	Human platelets	4 G	1983	1.6
TGF-β	Fsev-tranformed rat cells	126 L cond.med.	1984	1.7
bFGF	Bovine pituitary	2.3 KG	1987	1093
bFGF	Bovine brain	5 KG	1987	40
aFGF	Bovine brain	5 KG	1987	700
TGF-α	Human melanoma line	136 L cond.med.	1982	1.5
Urogastrone	Human urine	1000 L	1974	<1000

Table 1. The quantities of starting material for various preparations of some growth factors is shown. This illustrates the variable but often very large amounts required for purification, providing an indication of the low concentrations of many of these factors. Cond.med. denotes cell culture conditioned medium.

2.2 Structural properties

What we now loosely refer to here as growth factors are proteins consisting of single or multiple subunits of the same or different polypeptide chains, which can be post-translationally modified. For example a number of factors are glycosylated; however detailed analysis of the carbohydrate composition of even some of the well characterized growth factors is incomplete, due to insufficient amounts of material available.

Growth factors are in general of fairly low molecular mass, and range from about 6,000 for EGF to 82,000 for hepatocyte growth factor. They may be synthesized as high molecular weight precursors, which are then cleaved by proteolysis to generate active species. Frequently both single chain and multiple chain growth factors are stabilized by disulphide bonds; as secreted proteins (i.e. extracellular) the redox state of their environment would favour disulphide formation. Factors are often found associated with other proteins, including binding proteins apparently involved in transport or processing of the factor.

2.3 Differences between growth factors and hormones

Growth factors differ in some important ways from classical hormones, although there is not always a clear distinction between the two. The term "tissue growth factors" is sometimes used to emphasize the fact that they are not generally stored in specialized glands, but may derive from widely distributed cells or tissues. This has made the identification of the functions of growth factors very difficult, since it is not possible to carry out the classical ablation experiments in which the synthesis or storage organ of a hormone is removed to determine the effects on the animal. As suggested above in connection with tumour growth factors, 'normal' physiological growth factors are likely to have autocrine, as well as paracrine and endocrine effects.

2.4 Growth factors can have a range of important biological activities

Since it has been very difficult to study the functions of growth factors with the techniques used for the well known hormones, these factors were initially characterized and studied on the basis of their activities *in vitro*, often involving proliferation assays using test cells of various kinds. However further studies have revealed that many well known growth factors have a range of activities not necessarily related to growth. These include for example chemotaxis, differentiation and cell survival, and effects such as vasodilation and inhibition of gastric acid secretion at the level of specific organs.

Despite the recognition of such a range of activities for some factors, their *in vivo* functions and mechanisms of action are still not clear. What is clear however, is that growth factor receptors are crucial in determining which cell types will respond, and just which of a possible range of responses will be triggered. All known growth factors have such specific and often very high affinity binding proteins present on the surfaces of responsive cells; some growth factors can have more than one receptor, while some receptors can bind more than one factor (see Chapter 3).

A particularly intriguing question arises once the factor has bound its specific receptor as to whether the information for the proliferation signal is then contained entirely within the receptor, whether the complex as such is important, or whether the growth factor itself needs to be internalized into the cell to trigger additional signals. These possibilities are not mutually exclusive, since the mitogenic stimulation of a cell clearly involves a sustained alteration to a very large number of intracellular events which result eventually in the cell dividing and growing. What is known is that there exists a hierarchy of control processes that is triggered following the initial binding of a growth factor to its receptor, the details of which are the subject of subsequent chapters.

Growth factors almost certainly play crucial roles in a wide variety of processes, as suggested for example from studies with PDGF-like proteins (Table 2), although it should be emphasized again that their actual functions are still very poorly understood. It is likely that growth factors are involved in development, normal cell and tissue maintenance, wound repair and disease, and it is possible that many have biologically important but as yet undiscovered activities.

An example of a recently recognized activity is afforded by nerve growth factor (NGF), which is best known for having a foetal or neonatal role important for the survival and differentiation of the sympathetic adrenergic system. This neurotrophic factor

PDGF-LIKE PROTEIN	SOURCE	POSSIBLE ROLE
PDGF BB	Cytotrophoblast in first trimester human placenta	Development
PDGF	Smooth muscle cells from new born rat aorta	Embryogenesis/ development
PDGF AA	Fibroblasts/smooth muscle cells from various species stimulated with interleukin-1	Cell/tissue maintenance
PDGF AB,BB,AA	Platelets from all species tested	Wound repair
PDGF AA,BB	Macrophages from various species	Inflammatory responses
PDGF	Arterial endothelial cells from various species	Vascular repair/ atherosclerosis
PDGF	Vascular smooth muscle cells from various species	Vascular repair/ atherosclerosis

Table 2. Possible biological roles of PDGF-like proteins. Although the actual functions of growth factors are still very poorly understood, these proteins probably play crucial roles in a wide variety of processes. PDGF-related proteins have been detected in many different cell types, and these examples illustrate the range of physiological situations in which they may be important.

was originally named for its ability to stimulate neurite extensions from neuronal cells, a form of growth not directly related to cellular division. However it has recently been discovered that NGF does indeed stimulate some haematopoietic progenitor cells to divide as well as influencing their differentiation, and so it can now be considered to live up to its name as a true growth factor.

3. Some of the main growth factor families

While there are a large number of growth factors known, with more being discovered every year, it is not necessary to cover them all in detail. Instead, some of the better known families of structurally related factors are considered, since the properties they exhibit form general principles applicable to other individual growth factors or their families.

3.1   The epidermal growth factor family

The members of this family of factors are linked through limited structural similarity (as illustrated in figure 3). However, biologically these factors are very similar and in view of their apparent interaction with a common receptor protein this is perhaps not surprising.

```
VVGF (1-37)         MSMKYLMLLFAAMIIRSFADSGNAIETTSPEITNATT
Amphiregulin (1-38) SVRVEQVVKPPQDKTESENTSDKPKRKKKGGKNGKKRR

                    * * *** *    ** **      *   * **
EGF (1-53)          NSYPGCPSSYDGYCLNGGVCMHIESLDSYTCNCVIGYSGDRCQTRDLRWWELR
Urogastrone (1-53)  NSDSECPLSHDGYCLHDGVCMYIEALDKYACNCVVGYIGERCQYRDLKWWELR
TGFα, human (1-50)  VVSFHNDCPDSHTQFCFH-GTCRFLVQEDKPACVCHSGYVGARCEHADLLA
VVGF (38-100)       DIPAIRLCGPEGDGYCLH-GDCIHARDIDGFYCRCSHGYTGIRCQHVVLVDYQRSENPNTTTS
Amphiregulin (39-84) NRKKKNPCNAEFQNFCIH-GECKYIEHLEAVTCKCQQEYFGERCGEK
```

Fig 3. A comparison of amino acid sequences of the EGF family. Amino acid sequences of the mature proteins for some members of the EGF family are shown with the characteristic cysteine motif residues underlined as are other invariant residues. The numbers of residues are indicated in brackets. Residues conserved across three or four members are indicated by an asterisk. The limited but clear structural relationships between EGF, Uro and TGFα, as well as the more distant similarities with VVGF and amphiregulin, are apparent.

Epidermal growth factor (EGF)

Mouse epidermal growth factor has a molecular mass of 6,000 and contains three intramolecular disulphide bonds which are essential for its mitogenic activity. It occurs in the mouse submaxillary gland as a 74,000 complex consisting of 2 EGF molecules bound to 2 binding proteins which have arginine peptidase activity, and are possibly involved in releasing active EGF from its high molecular weight precursor under the appropriate circumstances. The complex can be dissociated at high or low pH. The active growth factor is initially synthesized as a large prepro-EGF membrane glycoprotein precursor which contains, as well as the sequence for the mature EGF protein, seven other EGF-like repeats. It is not clear whether any of these repeat sequences form biologically active species.

EGF has a specific cell surface high affinity receptor, which is a 170,000 molecular mass glycoprotein, phosphorylated on tyrosines following ligand binding. The discovery that the oncogene *erbB* represents a truncated form of the EGF receptor has provided strong evidence supporting the link between oncogenes and cancer, via the aberrant stimulation of normal growth pathways (see Chapter 3).

Urogastrone (Uro)

A factor from the urine of pregnant patients was purified based upon its ability to inhibit gastric acid secretion; this factor was named urogastrone. Surprisingly, amino acid sequence data revealed that this protein was the human homologue of mouse EGF. Of the 53 amino acids of EGF and urogastrone, 37 are identical, with the essential disulphide bonds located in exactly the same positions. A higher molecular weight form of Uro (of about 30,000) can also be detected, and similar molecules are found throughout the body (Table 3). The biological properties of this factor are so far indistinguishable from EGF, so that the abbreviation EGF/Uro is often used.

EGF-RELATED MOLECULES FOUND IN HUMAN TISSUE FLUIDS

TISSUE FLUID	CONCENTRATION, ng/ml
Serum	1.0
Plasma	0.05
Milk	80
Urine	100
Gastric secretions	0.4
Saliva	2.7
Tears	5.1
Pancreatic secretions	8.5
Duodenal secretions	21

Table 3. EGF-related molecules found in human tissue fluids. The reliable detection of EGF and related molecules in the body was not possible until the development of sensitive ELISA assays, utilizing specific antibody reagents. When these became available, it was unexpectedly discovered that EGF-related molecules are quite widely distributed. Their different levels of expression in a range of human tissue fluids indicate a diversity of specific biological roles.

Transforming growth factor α (TGFα)

Originally this factor was isolated as a mixture of TGFα and transforming growth factor β from virally transformed cells and was termed sarcoma growth factor. TGFα has now been purified to homogeneity from a range of cells and tissues of normal and neoplastic origin. The original purification assay measured the ability of the factor to stimulate growth in soft agar (anchorage independence) of certain indicator cell lines, such as the NRK (normal rat kidney) cell line. With our present understanding of function, it is now more usual to assay the EGF-like mitogenic and receptor binding activities of TGFα.

This factor has 50 amino acids, and the human tumour-derived, and rat and mouse fibroblast-derived forms have 92-100% homology. However these molecules show only limited homology to other members of the EGF family; despite some 33-44% homology including conserved disulphides, there is no immunological similarity.

TGFα and EGF probably evolved from a common ancestral molecule. There are striking differences between the precursor polypeptide of EGF, which is derived from an unexpectedly large 1200 amino acid precursor, and TGFα which has a 160 amino acid precursor. TGFα interacts with the EGF receptor, and so far no other receptor specific for TGFα has been identified.

Vaccinia virus growth factor (VVGF)

Studies on vaccinia virus genes revealed the presence of EGF-like sequences, raising the interesting possibility that vaccinia virus infected cells could produce a protein with properties similar to EGF. The protein precursor molecule predicted on the basis of the gene sequence was 140 amino acids in size, of which 50 were homologous to the EGF family, with the disulphides in the same positions. Following searches for EGF-like activity in conditioned media from infected cells, VVGF was detected and purified to homogeneity.

This molecule is a novel member of the EGF family, which is unusual in that (i) it is larger than most other members (77 amino acids, molecular mass 9,000) (ii) it is glycosylated and (iii) it can be recognized by antibodies to EGF, but not by antibodies to Uro or TGFα. VVGF binds to the EGF receptor to stimulate receptor phosphorylation, and is a potent EGF-like mitogen. Sequences encoding similar factors are present in the related poxviruses, myxoma and Shope fibroma. The discovery of this poxvirus growth factor class may establish a new principal by which the acquisition or evolution of a growth factor gene may give a selective advantage to the efficient infection and/or reproduction of viruses.

Amphiregulin

Recently a new member of the EGF family, amphiregulin, was isolated from medium conditioned by a human breast carcinoma cell line. This factor (molecular mass 9,700) and an amino-terminal 6 amino acid truncated for, both show EGF-like activity in receptor competition and growth factor assays. The new factor is apparently intermediate in structure between the EGF/Uro molecules and the pox virus factors. Its general distribution remains to be determined.

3.2 Transforming growth factor β (TGFβ) family

Originally isolated as a transforming growth factor activity (see above for TGFα), TGF β was identified as forming one part of this activity. This factor was found to require the presence of the other component (TGFα) and was assayed for its ability to stimulate growth in soft agar, in the presence of other growth factors such as EGF or TGFα.

Serum contains high concentrations of TGFβ, originating from platelets; these cells have 100 times higher TGFβ specific activity than that of any other tissue. The factor has also been isolated from human placenta and bovine kidney, and has been detected in a variety of cell types. A new member of the family, now called TGFβ2, has 71% amino acid sequence homology with the initially isolated TGFβ (now termed TGFβ1), and was originally identified in porcine platelets. Molecular biological approaches have identified three additional highly related molecules, TGFβ3, TGFβ4 and TGFβ5. Little is known about whether these factors all have similar activities, although there is increasing evidence that they are functionally distinct.

TGFβ polypeptides are disulphide crosslinked dimers of two 12,500 molecular mass chains (schematic structures are shown in figure 4). Homodimers and heterodimers

A

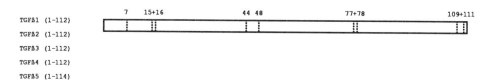

B

TGFß type: percentage identity

	1	2	3	4	5
1	100				
2	71	100			
3	72	76	100		
4	82	64	71	100	
5	76	66	69	72	100

Fig 4. Some members of the TGFβ family. The TGFs β 1-5 form a group of homologous proteins whose sequences from the data presently available are highly conserved across species barriers. (A) A common structure for the mature proteins for human TGFsβ1, β2 and β3, chicken TGFβ4 and amphibian TGFβ5 is shown, with the number of residues indicated in brackets. They share common cysteine residues, shown by dotted lines with the corresponding residue number; (B) Percent identies of TGFs β 1-5 as above. Data from *Kondaiah et al. J. Biol.Chem. 265, 1089-1093 (1990).*

of the β1 and β2 chains have been detected in porcine platelets, although the homodimeric β1 form seems to be most common.

Three TGFβ receptor types have been identified. Receptor Types I and II have high affinity for TGFβ1, with a lower affinity for TGFβ2, while Type III has high affinity for TGFsβ 1, 2 and 3. In general TGFsβ show a bewildering range of biological activities (Table 4), and can be mitogenic, or antiproliferative, depending on the cell type. They can also show complex effects in association with other growth factors. The TGFsβ may be important multifunctional regulators in many aspects of cell differentiation, growth, and general physiology. Other related members of the TGFβ family showing lesser degrees of homology to the TGFsβ, include mammalian Müllerian inhibitory substance, the inhibins and activins, the *Drosophila* decapentaplegic complex protein, the *Xenopus* vg1 gene product and bone morphogenetic proteins.

CELL TYPE	ACTIVITY
Epithelial cells Embryonic fibroblasts Endothelial cells	Inhibits proliferation
Osteoblasts Schwann cells	Stimulates proliferation
Fibroblasts	Stimulates extracellular matrix formation
Fibroblasts Monocytes	Stimulates chemotaxis
Capillary endothelial cells	Stimulates angiogenesis

Table 4. Some biological activities of the TGFβ family, together with the responding cell types are indicated. Most of these examples are from *in vitro* studies.

3.3 Insulin-like growth factor (IGF) family

There are three well established members of this family: insulin, IGF-I and IGF-II. (structures are illustrated in figure 5). Although insulin is 43% homologous to IGF-I (originally called Somatomedin C) and 41% homologous to IGF-II (originally called multiplication stimulating activity), it does not crossreact immunologically with the IGFs. Also unlike insulin, the IGFs circulate in plasma bound to specific carrier proteins. In the rat, the adult animal has IGF-I, while the foetus produces only IGF-II. IGFs are thought to be involved in mediating the effects of Growth Hormone, and their biological effects are qualitatively similar.

There are two classes of moderately specific high affinity glycoprotein receptors showing ligand-dependent tyrosine phosphorylation. The first is insulin specific, with weak binding of IGFs, and the second is relatively IGF-I specific, with binding of IGF-II and weak binding of insulin. A third IGF-II binding protein, the IGF-II receptor, binds IGF-I weakly, and does not bind insulin. This receptor is also the mannose-6-phosphate receptor and has been implicated in intracellular transport. The nature of the specificity of IGF-II binding to this receptor is not clear (see Chapter 3).

3.4 Fibroblast growth factor (FGF) family

The multiple members of this family were not recognized until recently, when it was

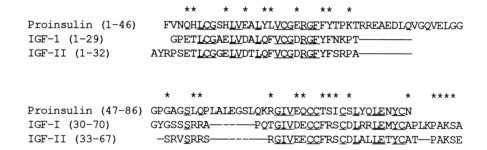

```
                            **   *  *  **   *  **  *
Proinsulin (1-46)    FVNQHLCGSHLVEALYLVCGERGFFYTPKTRREAEDLQVGQVELGG
IGF-1 (1-29)            GPETLCGAELVDALQFVCGDRGFYFNKPT————
IGF-II (1-32)        AYRPSETLCGGELVDTLQFVCGDRGFYFSRPA————


                         *   **       *  **  *** *      *    ****
Proinsulin (47-86)   GPGAGSLQPLALEGSLQKRGIVEQCCTSICSLYQLENYCN
IGF-I (30-70)           GYGSSSRRA——————PQTGIVDECCFRSCDLRRLEMYCAPLKPAKSA
IGF-II (33-67)          —SRVSRRS——————RGIVEECCFRSCDLALLETYCAT—PAKSE
```

Fig 5. Members of the IGF family. Amino acid sequences of the mature protein of human proinsulin, IGF-I and IGF-II, are shown with the number of residues indicated in brackets. Residues conserved across two members are indicated by an asterisk, with the characteristic cysteine motif residues underlined, as are other invariant residues.

discovered that a large number of apparently unrelated factors could be distinguished from most other growth factors on the basis of their unusual property of binding very tightly to heparin (a mixture of glycosaminoglycan molecules containing polysulphated polysaccharide chains). Not only was this property useful as a diagnostic test for the members of the family, but through this avid interaction heparin columns have also been exploited to give spectacularly good purification steps. These tremendously useful observations clearly divided FGFs into an acidic group (aFGF), which can be eluted from Heparin-Sepharose columns at about 1.0 M NaCl, and a basic group (bFGF), which elute at about 1.6 M NaCl.

FGFs stimulate mitogenesis in a wide variety of cells of mesodermal and neuroectodermal origin, have angiogenic properties, and may function as inductive agents during embryonic development.

Receptors for acidic and basic FGFs have been detected on the surface of most mesodermal and neuroectodermal cells investigated. They are approximately 125-165,000 in molecular mass, and their binding affinity for FGF varies with the cell type. It is not yet clear whether there are specific acidic and basic receptor types, or whether there is a single species of receptor which binds both FGF classes. A cDNA encoding a 130,00 molecular mass bFGF receptor from chicken has been isolated and found to encode a transmembrane protein with an intracellular tyrosine kinase domain (see Chapter 3).

Four additional members of what is now called the FGF superfamily were recently recognized following the sequencing of new oncogenes; these are discussed further below. The presently known members of the aFGF and bFGF family established initially from protein purification studies are shown in Table 5.

ACIDIC FIBROBLAST GROWTH FACTORS	ABBREV	SOURCE
Acidic brain fibroblast growth factor	BRAIN aFGF	human brain
Acidic brain fibroblast growth factor	BRAIN aFGF	bovine brain
Astroglial growth factor -1	AGF-1	bovine brain
Retina-derived growth factor	RDGF	bovine retina
Eye -derived growth factor-2	EDGF-2	bovine retina
Endothelial cell growth factor	ECGF	bovine hypothalamus
Endothelial cell growth factor	ECGF	bovine brain
Endothelial cell growth factor	ECGF	human brain
Heparin-binding growth factor α	HGFα	bovine brain
Heparin-binding growth factor β	HGFβ	bovine brain

BASIC FIBROBLAST GROWTH FACTORS	ABBREV	SOURCE
Basic brain fibroblast growth factor	bFGF	brain
Basic pituitary fibroblast growth factor	bFGF	pituitary
Cartilage-derived growth factor	CDGF	bovine
Chondrosarcoma-derived growth factor	ChDGF	rat chondrosarcoma
Eye-derived growth factor-1	EDGF-1	bovine retina
Astroglial growth factor-2	AGF-2	bovine brain

Table 5. Members of the aFGF and bFGF families. The presently known members of the aFGF and bFGF families established initially from protein purification studies, together with their sources of isolation and trivial names.

3.5 Haematopoietic growth factors

These important factors are involved in the control of haematopoiesis, the process in which eight types of mature blood cells are continuously produced from progenitor cells in the bone marrow via effects on proliferation, survival, differentiation and development. Eleven distinct factors have so far been characterized, their confusing range of names partly reflecting their multiple overlapping activities (Table 6). They are all glycoproteins (except Interleukin-1), with molecular masses of 15,000-20,000. The extremely complex interactions now being unravelled between these factors and the members of the eight blood cell lineages give fascinating glimpses into the way in which growth factors in general may interact with other cells, factors, and their receptors. For example, *in vitro* studies show that both GM-CSF and IL-3 can stimulate most granulocyte-macrophage progenitor cells to form maturing granulocytes and macrophages. However M-CSF stimulates these progenitors to form mainly macrophages, whilst G-CSF stimulates only some of the progenitors to

GROWTH FACTOR	ABBREV	SOME OTHER NAMES	ABBREV
Granulocyte-macrophage colony stimulating factor	GM-CSF		
		Colony stimulating factor-2	CSF-2
		Macrophage/granulocyte inducer-granulocyte-macrophage specific	MGI-IGM
		Human colony stimulating factor-β	HUMAN CSF-β
		Human pluripoietin-α	
		Human T-cell derived neutrophil migration-inhibition factor	NIF-T
Macrophage colony stimulating factor	M-CSF		
		Colony stimulating factor-1	CSF-1
		L-cell colony stimulating factor	L-CELL CSF
		Macrophage/granulocyte inducer-macrophage specific	MGI-IM
		Colony stimulating activity	CSA
Granulocyte colony stimulating factor	G-CSF		
		Macrophage/granulocyte inducer-2	MGI-2
		Leukaemia cell-differentiation-inducing factor	D-FACTOR
		Granulocyte/macrophage and leukaemic cell differentiation-inducing factor	GM-DF
		Human colony stimulating factor-β	HUMAN CSF-β
		Human pluripoietin	
Erythropoietin	EPO		
Interleukin-1	IL-1		
		Lymphocyte activating factor	LAF
		Mitogenic protein	MP
		Helper peak-1	HP-1
		T-cell replacing factor III	TRF III
		T-cell replacing factor	TR
		B-cell activating factor	BAF
		B-cell differentiation factor	BDF
		Endogenous pyrogen	EP
Interleukin-2	IL-2		
Interleukin-3	IL-3		
		Burst-promoting activity	BPA
		Haematopoietic cell growth factor	HCGF
		Multipotential colony stimulating factor	MULTI-CSF
		Mast cell growth factor	MCGF
		P-cell stimulating factor	PCSF
		Colony forming unit-stimulating activity	CFU-S
		Walter and Elisa Hall Institute-3 factor	WEHI-3 FACTOR
		Multicolony stimulating factor	
		P-cell stimulating factor	PSF
		Pan specific hematopoietin	PSH
		Histamine producing cell stimulating factor	
Interleukin-4	IL-4		
		B-cell growth factor	BCGF
		B-cell stimulating factor-1	BSF-1
		T-cell growth factor-2	TCGF-2
		Mast cell growth factor-2	MCGF-2
Interleukin-5	IL-5		
		Eosinophil differentiation factor	EDF
		Eosinophil colony stimulating factor	EO-CSF
		T-cell replacing factor	TRF
		B-cell growth factor II	BCGF II
		IgA-enhancing factor	IgA-EF
Interleukin-6	IL-6		
		Monocyte-derived human B-cell growth factor	
		Interferon-β2	IFN-β2
		B-cell stimulating factor-2	BSF-2
		Plasmacytoma growth factor	PCT-GF
		Interleukin hybridoma/plasmacytoma-1	IL-HP-1
		Hepatocyte stimulating factor	H
		Leukaemia inhibitory factor	LIF

Table 6. Haematopoietic growth factors. Commonly used names, abbreviations, and alternative names and abbreviations for the haematopoietic growth factor family. The wide range of names reflect the multiple and overlapping activities of many of these factors.

form mainly granulocytes. As the cells mature, they continue to express CSF receptors and are influenced further by the CSFs, whose concentrations also affect the processes of proliferation and differentiation.

3.6 Platelet-derived growth factor (PDGF) family

As described above, PDGF was discovered through the study of the growth stimulatory effects of blood serum as compared with plasma on fibroblasts in culture. PDGF is commonly isolated from outdated human platelet-rich plasma (produced for clinical use) after which it shows a range of molecular weights, due to proteolysis. Human PDGF is a glycoprotein, consisting of disulphide crosslinked dimers containing A-chains (11-12,000) and B-chains (12,000); these are mainly in the form of A-B dimers with some A-A and B-B forms. Interestingly, porcine PDGF is a B-chain homodimer with amino acid sequence more than 90% homologous to the human B-chain. No A-chain is apparently present in porcine platelets and the B-B PDGF homodimer exhibits the same range of activities as human PDGF, suggesting that the A-chain is not essential for the presently known PDGF biological activities.

Interest in the possible importance of A-chain molecules has recently been revived by the discovery of two types of PDGF receptors, one B-chain specific, and one A- and B-chain specific. The expression of these two receptor types as well as the chain compositions of the PDGFs themselves are independently regulated, suggesting a complex and highly controlled system for PDGF and receptor interactions, so generating specific biological responses.

A range of PDGF-like proteins have been detected in a variety of sources other than platelets (Table 2), reinforcing the notion that these molecules are important in many cellular processes.

4. Growth factors and oncogenes

4.1 PDGF and the *sis* oncogene

The first direct connection between a cancer-causing oncogene protein and a known protein expressed in normal cells was made in the case of PDGF B-chain and *sis*, the transforming protein of simian sarcoma virus (SSV). Following comparisons of the amino acid sequences of the two proteins, it was discovered that they were almost exactly the same (figure 6). This discovery was of great interest to cancer biologists because it dramatically reinforced the autocrine hypothesis from an unexpected direction, and strongly suggested that other oncogene proteins could be involved in growth control. The recognition that the SSV oncogene encodes a protein almost identical to the PDGF B-chain indicated that the tumourigenic properties of the virus are directly related to the production of a PDGF-like protein. Indeed, PDGF-like proteins, likely to be B-chain homodimers (like porcine PDGF), are produced by SSV-transformed cells. However the situation may be more complex than that envisaged by the minimal autocrine hypothesis, since there is some evidence for

```
                              ••
PDGF  B   (1-54)    SLGSLTIAEPAMIAECKTATEUFEISAALIDATNANFLUUPPCUEUQACSGCCN
sis      (67-120)   SLGSLSUAEPAMIAECKTATEUFEISAALIDATNANFLUUPPCUEUQACSGCCN


                                                                •        •
PDGF  B   (55-10    NANUQCAPTQUQLAPUQUAKIEIUAKKPIFKKATUTLEDHLACKCETUAAAAPU
sis      (121-174)  NANUQCAPTQUQLAPUQUAKIEIUAKKPIFKKATUTLEDHLACKCEIUAAAAU


                     •             ••                     •
PDGF  B  (109-160)  TASPGGSQEQAAKTPQTAUTIATUAUAAPPKGKHAKFKHTHDKTALKETLGA
sis      (175-226)  TASPGTSQEQAAKTTQSAUTIATUAUAAPPKGKHAKCKHTHDKTALKETLGA
```

Fig 6. PDGF and the *sis* oncogene. A comparison of the amino acid sequences of the mature B-chain of human PDGF and the corresponding region of the *sis* oncogene protein of simian sarcoma virus. The sequences are 95% identical, with the only differences (indicated as •) being compatible with slight sequence differences expected between highly conserved proteins of humans, and the Woolly Monkey from which the virus was originally isolated.

intracellular interactions between factor and receptor; following receptor activation other factors may be secreted.

4.2 FGFs and the *int-2*, *hst*/KS3, FGF-5 and FGF-6 oncogenes

Striking associations have now been made between the FGFs and four distinct oncogenes on the basis of sequence similarities (Table 7). The first was with the *int-2* oncogene, which was identified as a transforming gene following the integration of mouse mammary tumour virus into the mouse DNA. The virus does not carry the oncogene in this case, but integrates at 'hot' spots in the genome and at one such frequent insertion site this leads to enhanced expression of the adjacent *int-2* gene.

Two other oncogenes were then recognized following transfection assays with tumour derived DNA. Thus using DNA from a human stomach tumour, a Kaposi's sarcoma tumour and from a human bladder carcinoma three genes were independently identified (*hst*, KS3 and FGF5); *hst* and KS3 were found to be identical. Another oncogene, FGF-6, was detected following a screen for *hst*-related genes. None of these oncogene proteins is apparently directly related to known members of the FGF family that were previously isolated as proteins. Nevertheless all of these molecules with related sequences form the members of an FGF superfamily, whose precise functional relationships have yet to be clarified.

A puzzling aspect of the gene sequences for aFGF and bFGF is the absence of conventional signal peptides, suggesting that these factors are not secreted from cells by classical routes if at all. Interestingly however, the *hst*/KS3 and FGF5 oncogenes encode proteins with signal peptides, and studies with the *int-2* encoded protein suggest that the gene has evolved

FGF-LIKE ONCOGENE PROTEIN	SOURCE
int-2	Mouse mammary tumour virus
hst KS3	Human stomach tumour Human Kaposi's angiosarcoma Human hepatomas Human colon tumour
FGF-5	Human bladder carcinoma Human bladder tumour Human hepatoma Human endometrial carcinoma
FGF-6	Human stomach tumour probe / mouse DNA

Table 7. FGF-related oncogene proteins. The four classes of FGF-related oncogene proteins, together with the source from which they were originally isolated/detected.

so that its products can have different intracellular fates. As in the case of the *sis* oncogene, transformation by these gene products may not require the secretion of the transforming protein.

3

Receptors

1. Cellular monitors of the extracellular environment

In the broadest terms, receptors may be defined as the cellular components that monitor the status of the extracellular environment. They are responsible for interacting with regulatory molecules from outside the cell and passing a biochemical signal into the cell which can be interpreted as a measure of the concentration of such molecules. The cell can then respond to this information in the appropriate way. A precise definition of a receptor is not easy to make, and is likely to include a great many diverse molecules. For the purposes of this chapter only proteinaceous cell surface receptors will be considered; nuclear receptors for steroids, retinoids and thyroid hormones will be discussed in Chapter 7, while glycolipids that may be defined as receptors are beyond the scope of this book.

Receptors are involved in a great many essential cellular functions. In animal cells they can have positive or negative effects on cellular growth and they can mediate immunological recognition, chemotaxis, neurological signalling and many other functions. Even the roles that receptors play in what might at first appear to be purely mechanical interactions such as cell-cell contact and cell attachment to the extracellular matrix are now understood to have profound effects on cellular growth. Receptors are not just restricted to metazoan eukaryotes: they have also been extensively studied in unicellular eukaryotes such as yeast and slime moulds and in prokaryotes. In this chapter most emphasis will be placed on the mammalian receptors for peptide growth factors and hormones. However, receptors for other signalling molecules will be considered also, both in vertebrates and lower eukaryotes.

Despite the great diversity of known receptors, it has become apparent that many of them can be grouped into families which have related structures and functions. A multitude of variations on a few fundamental themes seem to be used in the transduction of extracellular signals into the cell. The structure and mechanism of action of receptors from a number of these families is now beginning to be understood. In this chapter, emerging themes in receptor structure and function will be discussed.

2. Receptors can possess an enzyme activity or directly regulate such an activity

2.1 Growth factor receptor tyrosine kinases

This family of growth factor receptors was among the first to be structurally characterized. The primary sequences show that these receptors possess an extracellular domain that is often rich in cysteines and N-linked glycosylation, and includes the site to which the growth factor or ligand binds. Each of these proteins also has a domain that is structurally related to the tyrosine kinase encoded by the *src* oncogene. These two domains are separated by a single stretch of twenty or so hydrophobic amino acids which serves to anchor the protein in the plasma membrane with the amino-terminal ligand binding domain outside the cell and the carboxy-terminal tyrosine kinase domain in the cytoplasm. The large size of the cytoplasmic domain (500-600 amino acids) alongside the single transmembrane span distinguishes these receptors from most others.

In the growth factor receptor tyrosine kinase family the binding of a growth factor to the extracellular domain of the receptor is translated into activation of the intracellular tyrosine kinase domain. The precise mechanism by which this signal is transferred from the extracellular to the intracellular domains is not yet clear, but it is likely that ligand induced receptor aggregation is important, at least for some members of the family. It is also conceivable that a conformational change could be passed directly through the transmembrane domain by a single receptor molecule or that another type of protein is also involved in the process. A longer term response to ligand binding is receptor internalization and degradation: this provides a means of attenuating the activation signal.

As is the case with *src* and other related oncogenes, an activated tyrosine kinase located at the cytoplasmic face of the plasma membrane has a strong growth stimulating effect on the cell. Several potentially important targets for growth factor receptors and other tyrosine kinases have recently been characterized, these include phosphatidyl inositol kinase (specific for the D-3 position of the inositol ring), phospholipase C-γ, the serine/threonine kinase p74^{raf} and GAP, the GTPase activating protein of p21^{ras}. These substrates will be considered in Chapters 4 and 5.

The growth factor receptor tyrosine kinase family can be divided into three principal subfamilies as shown in figure 1. These are discussed separately below.

The EGF receptor subfamily

In vertebrates this group of receptor tyrosine kinases consists of the receptor for epidermal growth factor and the related protein termed HER2/*neu*. In invertebrates such as *Drosophila* only one such protein has been characterized: it has been named DER (*Drosophila* EGF Receptor) and is as closely related to the human HER2/*neu* as it is to the human EGF receptor. The EGF receptor is encoded by the proto-oncogene *c-erbB* while HER2/*neu* is

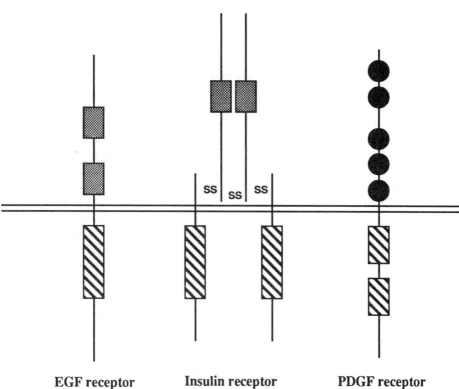

Fig 1. Structures of the receptor tyrosine kinases. The hatched boxes represent the cytoplasmic tyrosine kinase domains, the grey boxes the cysteine rich extracellular domains and the black circles the immunoglobulin-like domains.

also known to be produced by a gene with oncogenic potential. The connection between growth factor receptor genes and oncogenes will be explored in section 5 of this chapter.

Proteins of the EGF receptor subfamily are monomeric, with single transmembrane spans. They possess two cysteine-rich sequence repeats in their extracellular domains; these cysteines are conserved between members of the subfamily whereas the rest of the extracellular domain is not highly conserved. This part of the EGF receptor contains the ligand binding site and also has a dozen sites of N-linked glycosylation.

The EGF receptor shows complex binding properties and appears to exist in a high and a low affinity binding state. Furthermore there exist a number of different ligands for the EGF receptor including epidermal growth factor itself and the related, but distinct, polypeptides TGFα (transforming growth factor α), VVGF (vaccinia virus growth factor)and amphiregulin (see also Chapter 2). These molecules all interact with the EGF receptor in very similar ways and apparently cause comparable effects on cells in culture. A ligand for HER2/*neu* has only recently been identified and is significantly larger than any of the EGF receptor ligands. The relationship of the HER2/neu ligand to the EGF family is not yet known.

Within the cytoplasmic domain of these receptors there are three identifiable regions. Just beyond the transmembrane domain is a region of about fifty amino acids (sometimes referred to as 'juxtamembrane') that is well conserved within each subfamily, but not between them. Threonine residue 654 of the EGF receptor, which is located in this area, is phosphorylated by protein kinase C: this modification contributes to the conversion of the receptor binding site from the high to the low affinity state. Within the cell this site is phosphorylated in response to any treatment that activates protein kinase C, such as growth factors that stimulate phosphatidyl inositol turnover. By this means the affinity of the EGF receptor for its ligand can be decreased by other growth stimuli: this phenomenon is referred to as 'transmodulation'. The physiological importance of transmodulation has not yet been determined but it could clearly act as a form of heterologous desensitization (see Chapter 8).

The next region of the cytoplasmic domain of the EGF receptor subfamily confers the tyrosine kinase activity. This part of the molecule is very strongly conserved within the whole family of tyrosine kinases, with the *src* oncogene product usually being considered as the prototype. It also shows homology to serine/threonine kinases and, to a lesser extent, to other nucleotide binding proteins such as the G proteins. A conserved lysine residue (EGF receptor residue 721) is essential for kinase function: mutation of this amino acid generates a receptor that is not only deficient in kinase activity but also devoid of biological function. Another essential region contains the motif Gly-X-Gly-X-X-Gly, located in the amino terminal part of the kinase domain: this appears in most nucleotide binding proteins and is involved in contact with the phosphoryl groups of the bound nucleotide.

Finally, the carboxyl-terminal region of the EGF receptor subfamily has about 200 amino acids that are not homologous to the *src* oncoprotein. This region contains the major tyrosine sites at which the EGF receptor phosphorylates itself upon binding EGF ("auto-phosphorylation"). The autophosphorylation sites are conserved in HER2/*neu* but not in receptor tyrosine kinases of other subfamilies. The precise role of autophosphorylation in the functioning of the EGF receptor has not been established. Receptors lacking these sites behave very similarly to normal receptors. *In vivo* and *in vitro*, autophosphorylation does not greatly affect kinase activity or ligand binding affinity; it seems likely that these phosphorylations have a relatively subtle regulatory role.

The means by which binding of ligand to the extracellular domain leads to stimulation of the tyrosine kinase activity of the intracellular domain is not yet completely understood. However for the EGF receptor it does appear that receptor dimerization occurs upon EGF binding and it is likely that this step is required for kinase activation (see figure 2).

The PDGF receptor subfamily

This subfamily consists of the two closely related receptors for platelet derived growth factor (type A binds the PDGF dimers AA, AB and BB, and type B binds only PDGF BB; see Chapter 2), the receptor for macrophage-colony stimulating factor (M-CSF, also named

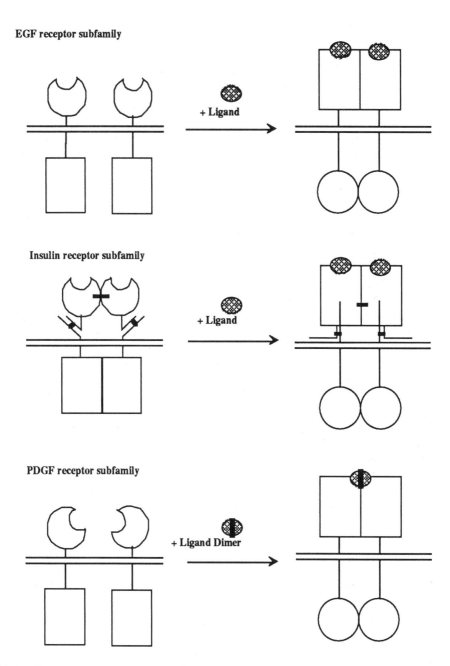

Fig. 2 Models of receptor activation by dimerization for the receptor tyrosine kinase subclasses. Activation could result from monomeric ligand binding causing a conformational change in the extracellular domain which leads to receptor dimerization and activation (EGF receptor), by monomeric ligand binding to dimeric receptor causing confirmational change within the complex (insulin receptor) or by dimeric ligand binding to two monomeric receptors to promote receptor dimer formation (PDGF receptor).

CSF-1), which is the product of the *c-fms* proto-oncogene, the receptor for basic fibroblast growth factor (the product of the *flg* gene) and the putative receptor encoded by the proto-oncogene *c-kit*. It is also likely that the partly characterized *c-ret* and *c-sea* proteins may be receptors belonging to this subfamily. Receptor tyrosine kinases of this subfamily have the same general topology as the EGF receptor, but have two clear differences. Firstly, they do not possess the two cysteine rich repeats found in the extracellular domain of the EGF receptor subfamily. Instead they have five β-sheet rich repeats of about 100 amino acids each of which are similar to motifs found in the immunoglobulin superfamily. Secondly, the PDGF receptor subfamily members have a unique region of about 100 amino acids known as the "kinase insert": this structure is found at precisely the same position in the middle of the conserved tyrosine kinase domain of each member of the subfamily, but the kinase inserts of the different proteins are not closely related. All of the receptors with kinase inserts also possess the five immunoglobulin-like domains; these two motifs thus provide hallmarks for this subfamily.

Like the EGF receptor subfamily, binding of the ligand PDGF to the PDGF receptor stimulates the activity of the intracellular tyrosine kinase domain. As a consequence a number of events are induced within the cell including receptor autophosphorylation, calcium fluxes, pH changes, phosphatidyl inositol hydrolysis and phosphatidyl inositol phosphorylation. Eventually receptor occupancy leads to DNA synthesis and cell division. The kinase insert of the PDGF receptor appears to be very important in receptor function: mutants lacking this region display all the early responses of the wild type receptor to PDGF with the exception of stimulation of the phosphatidyl inositol kinase. However, the kinase insert deletion mutants are only poorly able to induce DNA synthesis and mitosis. This indicates that the phosphatidyl inositol kinase activity (directed towards the D-3 position of the inositol ring) may be essential for the mitogenic response. Interestingly, the kinase insert mutants fail to interact with phosphatidyl inositol kinase, p74^{raf} or GAP raising the possibility that the kinase insert may be some sort of recognition or protein-protein binding domain. However it should be added that receptors without a kinase domain insert (e.g. EGF receptor) can interact with members of this group of proteins.

The insulin receptor subfamily

The insulin receptor subfamily of receptor tyrosine kinases consists of the insulin receptor itself (mammalian and dipteran), the receptor for the insulin-like growth factor type I (IGF-I) *c-ros*, and the incompletely characterized products of the proto-oncogenes *c-trk* and *c-met*. These receptors differ from those discussed above in a number of ways. Perhaps the most striking is the fact that they are heterotetrameric, comprising two α subunits and two β subunits that are covalently linked by disulphide bonds. The receptor is the product of a single gene that encodes an αβ precursor which is proteolytically cleaved to give the mature subunits. The α subunit is entirely extracellular and contains one cysteine rich domain (two per receptor) that is similar to those found in the EGF receptor. Each α subunit is linked to the other α subunit and also to one of the β subunits by disulphide bonds. The ligand binding ability appears to reside principally in the α subunits. The β subunits contain the trans-membrane domains (one each) and also the tyrosine kinase domains; these contain no kinase inserts.

Like other receptor tyrosine kinases, the binding of ligand to the insulin receptor results in stimulation of the activity of the intracellular tyrosine kinase domain, with one of the major substrates being the receptor itself. Autophosphorylation occurs at several tyrosine residues in the β-subunit. It has been established for the insulin receptor subfamily, but not for the other subfamilies of receptor tyrosine kinases, that autophosphorylation leads to activation of the kinase activity. The residues that appear to be most important in this regulation are the neighbouring tyrosines 1150 and 1151, which are located in a position analagous to the major autophosphorylation site of pp60^{v-src} in its kinase domain. This situation is different from the EGF receptor where autophosphorylation occurs outside the kinase domain close to the carboxy-terminus.

The physiological effects of insulin upon cells are very varied and complex: in part this is due to the fact that insulin binds both to the insulin receptor and, with lower affinity, to the closely related IGF-I receptor. IGF-I itself also binds to both receptors and also to another receptor, the IGF-II receptor; this is not a tyrosine kinase (see section 3.3 of this chapter). A third growth factor of the insulin family, IGF-II, binds most efficiently to the IGF-II receptor but also to the IGF-I and insulin receptors. The exact role of each receptor in mediating the response to each of these growth factors is still under investigation.

While many of the early effects of insulin receptor activation on the cell are similar to the activation of other receptor tyrosine kinases, there is one immediate response that appears to be unique. This is the activation of a form of phospholipase C that is specific for a glycosyl-phosphatidyl inositol. The resulting hydrolysis reaction yields diacylglycerol, which may activate protein kinase C, and an inositol phosphate-glycan, which may also be a second messenger. The means by which the insulin receptor controls this pathway are not understood, although there has been speculation that a GTP binding protein is involved; it is known that receptor tyrosine kinase activity is necessary.

Receptors with separate cytoplasmic tyrosine kinase subunits

A great many tyrosine kinases do not have hydrophobic transmembrane domains: these proteins reside in the cytoplasm, either free or attached to the inner leaflet of the plasma membrane by myristylation. It has been argued that these kinases could form part of a receptor complex in which the kinase containing subunit (comparable to the β- subunit of the insulin receptor) is entirely cytoplasmic and not covalently linked to the other subunit which would be extracellular and membrane spanning. Recently evidence has emerged to support this model in one system at least. In T-lymphocytes the pp60^{src} related cytoplasmic protein tyrosine kinase p56^{lck} has been found to be functionally and physically associated with the cell surface glycoproteins CD4 and CD8. These proteins act as receptors for major histocompatibility antigens and are important accessory molecules in the T-cell receptor recognition of antigen bearing cells (see Figure 3). It is likely that the interaction of CD4 or CD8 with major histocompatibility complex determinants causes activation of the p56^{lck}

associated with their short cytoplasmic domains (see Chapter 5). This leads to tyrosine phosphorylation of the ζ (zeta) chain of the antigen receptor (CD3).

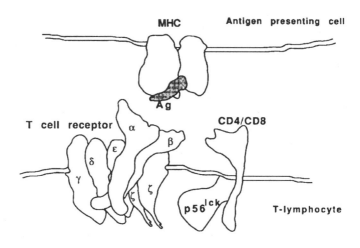

Fig 3. Model of T-cell antigen receptor showing its interaction with the p56^{lck}/CD4 or CD8 complex and major histocompatibility antigen (MHC) on the antigen (Ag) presenting cell.

2.2 Receptors regulating enzymes via the action of GTP-binding proteins

A large number of receptors for a wide range of signalling molecules have been found to control the activity of intracellular enzymes by interacting with GTP-binding proteins at the inner face of the plasma membrane. This signal transduction mechanism is discussed in detail in Chapter 4. The pathway can be summarized as follows: binding of ligand to the extracellular part of the receptor causes a conformational change to be transmitted to the cytoplasmic portion of the receptor. This conformational change alters the interaction of the receptor with a heterotrimeric guanine nucleotide binding protein in the cytoplasm. These proteins, known as "G proteins", consist of an α subunit of molecular mass between 39,000 and 54,000, which possesses the nucleotide binding site, a β subunit and a γ subunit of molecular mass 35-36,000 and about 8,000 respectively. Interaction of the G protein with the activated receptor changes its own activation state, typically from an inactive, GDP bound form to an active GTP bound form. This is achieved by stimulating the rate at which the nucleotide exchanges on and off the G protein: since the form of guanine nucleotide in the cytosol is predominantly GTP rather than GDP, this leads to an accumulation of GTP on the G protein. The activated G protein then interacts with an intracellular effector enzyme resulting in an alteration of the enzymic activity of this target. The effect of the activated G protein is attenuated by its intrinsic hydrolytic activity which causes the degradation of bound GTP to GDP. A second attenuation mechanism that is characteristic of these types of receptors is that they undergo a decrease in affinity for their ligand when they interact with activated (GTP bound) G proteins.

The details of these signal transduction pathways vary considerably from system to system but all have the same basic theme. In particular, the cell surface receptors which control these G proteins have a clearly recognizable structure: although their amino acid sequences are usually not highly conserved, they all possess seven highly hydrophobic regions. These regions typically consist of about twenty amino acids of a predominantly hydrophobic nature. The first protein to be characterized that showed this feature was bacteriorhodopsin, the protein of the purple membrane of the archaebacterium *Halobacterium halobium* which uses the energy of visible light to power a proton pump. Electron scattering techniques were used to obtain a low resolution structure for this protein in two-dimensional crystals in the purple membrane: it was seen to span the lipid bilayer seven times. Later, when the primary structure of the mammalian light transducing protein rhodopsin was established, this motif was seen again and a seven membrane spanning structure suggested (see figure 6). While there is no direct evidence that the mammalian G protein coupled receptors actually adopt this structure, it appears to be a reasonable assumption.

Some of the G protein coupled, seven membrane spanning receptors have now been studied in very great detail, particularly the adrenergic receptors and the rhodopsins. The adrenergic receptors have been defined using several of the general strategies outlined in figure 4. This group of receptors will be discussed below in some detail as a paradigm for the whole family.

Adrenergic receptors

Several distinct subtypes of receptors for catecholamines have been defined on the basis of their pharmacological specificity and physiological actions. The different types of receptors are coupled to different enzyme systems: β_1- and β_2-adrenergic receptors act through the G protein G_s to stimulate adenylyl cyclase leading to increased cytoplasmic levels of the second messenger cyclic AMP. α_2-adrenergic receptors oppose the effects of the β-receptors by acting through the G protein G_i which inhibits adenylyl cyclase. α_1-adrenergic receptors activate phosphatidyl inositol breakdown by phospholipase C leading to the production of the second messengers diacylglycerol, which activates protein kinase C, and inositol trisphosphate, which causes release of calcium from intracellular stores, (the G protein that mediates this response, often referred to as G_p, is not yet well characterized). Other less well studied subtypes also exist.

The receptors of this subfamily all possess seven highly hydrophobic "transmembrane spans". These hydrophobic regions are the most highly conserved between the different receptors (71% between human β_1- and β_2-receptors). The loops of hydrophilic amino acids that are presumed to extend into the extracellular space or the cytoplasm are much less conserved (43% between human β_1- and β_2-receptors). The membrane spanning regions appear to be involved in much more than simply locating the receptor in the lipid bilayer. Extensive site-directed mutagenesis work and creation of chimeric adrenergic receptors has shown that these hydrophobic regions are essential for the binding of the ligand, whereas the hydrophilic loops are not. Several of the hydrophobic spans are involved in ligand binding, most importantly numbers II, III and VII (see figure 5).

The adrenergic receptors have four extracellular regions: the amino-terminus and three

1. Purification and sequencing of receptor proteins

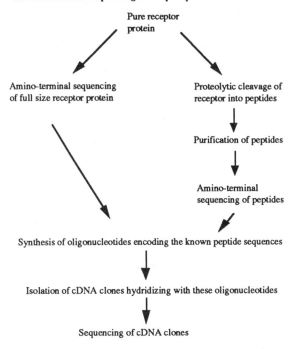

Pure receptor
protein

Amino-terminal sequencing
of full size receptor protein

Proteolytic cleavage of
receptor into peptides

Purification of peptides

Amino-terminal
sequencing of peptides

Synthesis of oligonucleotides encoding the known peptide sequences

Isolation of cDNA clones hydridizing with these oligonucleotides

Sequencing of cDNA clones

2. Expression cloning of receptor proteins

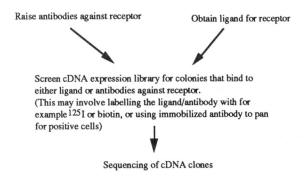

Raise antibodies against receptor

Obtain ligand for receptor

Screen cDNA expression library for colonies that bind to
either ligand or antibodies against receptor.
(This may involve labelling the ligand/antibody with for
example ^{125}I or biotin, or using immobilized antibody to pan
for positive cells)

Sequencing of cDNA clones

3. Cloning of receptor proteins by homology

Screen cDNA libraries at low stringency for hybridization to
probes based on the known sequence of characterized receptors
OR
Use the polymerase chain reaction with primers based on the
sequence of regions of maximum conservation in the receptor family

Fig 4. Strategies for molecular cloning of receptors.

external hydrophilic loops. These vary greatly in size and sequence between different members of the family. One common feature is that most of these receptors have at least one potential site of N-linked glycosylation close to the amino-terminus.

The intracellular regions are somewhat better characterized. There are four of these: three cytoplasmic loops and the carboxy-terminal tail. It is likely that these regions are involved in the interaction of the receptors with the G proteins and hence may determine the specificity of the coupling. The first two cytoplasmic loops are more conserved than the third loop and the carboxy-terminal tail. Deletion studies suggest that the third cytoplasmic loop is essential for the interaction of the receptors with G proteins.

A focus of considerable interest in the study of adrenergic receptors has been regulatory modifications that lead to their "desensitization". It has long been known that cells stimulated with a particular adrenergic agonist soon become refractile to further stimulation by the same agonist and sometimes other catecholamines. The process by which the cell becomes refractile to further stimulation by a single particular agonist is termed homologous desensitization. The β_2-adrenergic receptor becomes phosphorylated soon after it binds to its ligand: this phosphorylation, which occurs close to the carboxy-terminus, uncouples the receptor from G_s and thus attenuates its stimulatory signal. The kinase responsible for this phosphorylation which has been extensively studied is referred to as the β-adrenergic receptor kinase (βARK) and is capable of phosphorylating only receptors that are bound to ligand and not unoccupied receptors. This kinase is capable of phosphorylating all the adrenergic receptor subtypes, in each case in a strictly ligand dependent manner, thereby attenuating the signal received by the cell.

A second form of signal damping, heterologous desensitization, occurs when the binding of ligand to one receptor subtype leads to a decrease in the ability of other receptor subtypes to transduce a stimulus into the cell. The mechanism for this attenuation also involves phosphorylation of receptors close to their carboxy-termini, but in this case in a ligand independent manner. A number of kinases may be involved, for example, activation of the α_1-adrenergic receptor by agonist binding leads to the activation of protein kinase C via the phophatidyl inositol breakdown pathway; protein kinase C then phosphorylates the β_2-adrenergic receptor and prevents it from responding to agonist. Similarly, activation of the β_2-adrenergic receptor leads to activation of the cAMP dependent protein kinase which then phosphorylates the α_1-adrenergic receptor, preventing it from signalling. Aspects of feedback controls are discussed further in Chapter 8.

Neurotransmitter and neuropeptide receptors

The brain and nervous system contain a great many specialized signalling systems, many of which involve G proteins controlled by receptors of the seven membrane span type. The best studied of these receptors are the muscarinic acetylcholine receptors. In humans these form a family of at least five receptors: M1, M2, M3, M4 and M5. Certain of these different forms were distinguished originally by their relative abilities to recognize the antagonists piperazine and AF-DX116; the primary structure of each has now been established. Their sequences are similar to each other by between 30 and 55%. Like the adrenergic receptors

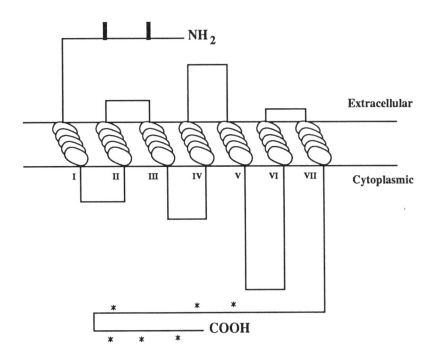

Fig 5. Structure of the G protein linked receptors using the adrenergic receptors as a model. Sites of phosphorylation are indicated (*). Glycosylation sites are marked by heavy lines. Roman numerals indicate the seven
transmembrane spans.

they are most conserved in their hydrophobic transmembrane spans. M3 is only slightly
more closely related to M4 than it is to the adrenergic receptors.

Like the adrenergic receptors, the muscarinic acetylcholine receptor subfamily couple to a
number of different signal transduction systems via different G proteins. M2 and M3
efficiently inhibit adenylyl cyclase by acting upon G_i, whereas M1 and M4 fail to inhibit
adenylyl cyclase but strongly stimulate phosphatidyl inositol turnover, presumably via the
putative phospholipase C controlling G protein, G_p. Interestingly, when M2 or M3 are
expressed in cells at a very high level they are able to weakly activate phosphatidyl inositol
breakdown, indicating that a degree of promiscuity exists in the system with receptors being
able to talk to the "wrong" G proteins under some circumstances. This failure of specificity
has been found to become much more significant in subcellular systems.

Muscarinic acetylcholine receptors have also been reported to control a potassium ion
channel through a G protein that has been named G_k. Some other neurotransmitter receptors
that have been shown to be G protein coupled and to possess seven hydrophobic regions
include the dopamine receptor family, of which D_2 activates G_i, the serotonin (5-hy-

channel through a G protein that has been named G_k. Some other neurotransmitter receptors that have been shown to be G protein coupled and to possess seven hydrophobic regions include the dopamine receptor family, of which D_2 activates G_i, the serotonin (5-hydroxytryptamine) receptor family, of which the 5-HT$_{1\beta}$ receptor activates G_i, the μ and δ opiate receptors and the adenosine receptor family. In addition the receptors for the neuropeptide substance K and the blood pressure controlling peptide angiotensin have been found to fall into this category: one of the angiotensin receptors appears to be the product of the proto-oncogene *c-mas* (see section 4 of this chapter).

As will be apparent from the above discussion, the G protein coupled, seven membrane spanning receptors are proving to be a very frequently used mechanism in many diverse systems. Indeed, certain growth factor receptors of unknown structure are known to have G protein mediated signal transduction pathways: it seems very likely that these will prove to have the characteristic structural motifs of the receptors described above.

Rhodopsins

While the proteins of the rhodopsin family may perhaps not be considered to be receptors in the conventional sense, they are clearly involved in a signal transduction pathway whereby an external stimulus (a photon of light) is translated into a cellular response (an action potential) via a second messenger (cyclic GMP; cGMP). In brief, the signalling cascade operates as follows: light incident upon the rod cells of the mammalian retina causes photoisomerization of retinal, the chromophore attached to opsin (the complex of the protein, opsin, plus bound retinal is referred to as rhodopsin). This isomerization triggers a conformational change in opsin which activates the associated G protein known as transducin. The GTP bound form of transducin then activates the effector enzyme cGMP phosphodiesterase which induces a rapid decrease in the cGMP levels in the rod cytosol. The outer segments of the rods contain cation-specific ion channels that are opened by binding cGMP. Reduction in the level of cGMP causes the channels to close: hyperpolarization of the membrane results which is transmitted to the synaptic terminal. A feedback mechanism acts whereby the decrease in intracellular calcium concentration due to the closing of the cation channels activates guanylyl cyclase which synthesizes new cGMP leading to a reopening of the ion channels. The system is exquisitely sensitive, with a large amplification of the signal at each step, and turns on and off extremely rapidly. Due to the abundance of the components in the retina the system has been studied in great detail.

The structure of rhodopsin itself is very similar to that of the adrenergic and related receptors in that it has seven hydrophobic transmembrane spans (see figure 6). The chromophore retinal, which can be considered to be the ligand, is attached by a Schiff's base linkage to lysine residue 296, which is in the middle of the seventh transmembrane α-helix. There are two sites of N-linked glycosylation near to the amino terminus of the protein (aspartates 2 and 15) and two sites of palmitylation near the carboxy-terminus (cysteines 322 and 323); these latter modifications would be expected to anchor the carboxy-terminal region to the cytoplasmic face of the rod outer disk membrane.

Like the adrenergic receptors, rhodopsin becomes phosphorylated following illumination. The kinase responsible, rhodopsin kinase, causes uncoupling of rhodopsin from transducin

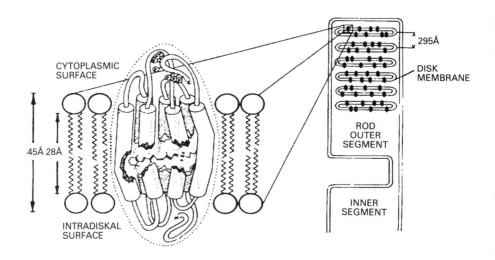

Fig 6. Model of rhodopsin showing it inserted in the membrane of the rod outer segment. The seven known phosphorylation sites are marked with the letter P. The retinal chromophore is shown surrounded by the seven membrane spanning α-helices (from *Dratz & Hargrave, TIBS, 8, 128*).

and is thus analagous to the β-adrenergic receptor kinase. Another protein, "arrestin", binds to the cytoplasmic loops of activated rhodopsin and promotes the uncoupling process. Both these mechanisms contribute to the exquisite sensitivity of the system.

The cone cells of the retina that are involved in colour vision contain different but related photoreceptors. There are three of these receptors, each absorbing light of different wavelengths. They all have a similar seven hydrophobic α-helix structure and appear to function in a similar manner to rhodopsin but couple via a distinct form of transducin.

2.3 Receptors with guanylyl cyclase activity

Like the well studied cyclic AMP, cyclic GMP (cGMP) is also a second messenger. It acts through a number of effectors such as the cGMP-dependent protein kinase and cGMP gated ion channels (see previous section). cGMP is generated from GTP by guanylyl cyclase: two forms of this enzyme exist, one cytosolic and the other membrane bound. Recently the membrane associated form of guanylyl cyclase was purified from sea urchin spermatazoa; this enzyme was known to be regulated by chemotactic peptides secreted by sea urchin eggs such as "speract" in the species *Strongylocentrus purpuratus* and "resact" in *Arbacia punctulata*. In both cases it was found that the membrane form of guanylyl cyclase copurified with the chemotactic peptide receptor. Biochemical characterization and molecular cloning of these proteins established that the cell surface receptor for resact was the same polypeptide as the high molecular weight form of the membrane guanylyl cyclase i.e. both these activities reside in the same molecule. The sequence shows that there is an extracellular portion of the protein that is heavily N-glycosylated, a single hydrophobic membrane spanning region and an intracellular domain that has a region with strong homology to the soluble form of guanylyl cyclase. The external sequences appear to be required for peptide binding while the intracellular domain is necessary for guanylyl cyclase activity. Somewhat surprisingly, the cytoplasmic part of the molecule also has a region of homology with the protein kinase family (this is not found in the soluble forms of the enzyme). By contrast, the receptor for speract does not possess an intrinsic guanylyl cyclase activity and it has only a small cytoplasmic domain. It presumably associates closely with a wholly intracellular form of guanylyl cyclase.

Another example of a receptor with guanylyl cyclase activity has now been established. The mammalian vasodilator atrial natriuretic peptide (ANP) binds to a 130,000 molecular mass receptor that is a membrane bound guanylyl cyclase. A second homologous brain form of this receptor has also been identified. It seems likely that more receptors of this kind will soon be discovered. Whether they form as large a group as the receptor tyrosine kinases or the G protein linked receptors remains to be seen.

2.4 Receptors with phosphatase activity

The action of protein kinases has long been recognized to be extremely important in the regulation of cellular processes. Section 2.1 of this chapter discussed at length the receptor tyrosine kinases and their roles in the control of cell growth. Phosphorylation is a key regulatory mechanism for many enzymes and while a great deal of work has been carried out on the kinases, much less is known about the protein phosphatases, the enzymes that remove phosphate groups from proteins by catalysing the hydrolysis reaction. Clearly phosphatases could be equally as important as kinases in regulating the phosphorylation state of target proteins.

While a number of phosphatases that remove phosphate from serine and threonine residues have been well characterized (see Chapter 6), it is only recently that tyrosine phosphatases have been purified and studied in detail. When primary sequence information was obtained from the major cytoplasmic protein tyrosine phosphatase of human placenta (known as 1B), a highly significant homology was found to a part of a cell surface molecule of haematopoietic cells, CD45 or the leukocyte common antigen. This protein, which is found on all cells of lymphoid and myeloid lineages, has three distinct domains. At the amino-terminal end there is a large, glycosylated, cysteine rich extracellular region of about 500 amino acids. There is then a single transmembrane domain of 22 hydrophobic amino acids, followed by a large intracellular sequence of some 700 amino acids. It is in this cytoplasmic region that the homology to the protein tyrosine phosphatase exists: it occurs as two tandem, highly homologous repeats which are both greater than 30 % identical to the sequence of protein tyrosine phosphatase 1B. It has now been shown directly that CD45 possesses tyrosine phosphatase activity (see Chapter 6).

The exact way in which CD45 functions is not yet completely clear. It is now apparent that there are several different isoforms of the molecule that are found in different cell types: the differences between them lie in their extracellular domains. There is only one gene for all the CD45 isoforms; the differences are the result of alternative splicing of the mRNA which results in a number of different possible insertions into the extracellular domain. It seems likely that the various isoforms (at least three) interact with different extracellular molecules resulting in the modulation of the phosphatase activity of the cytoplasmic domain. The "ligands" for the CD45 proteins have not yet been identified.

In non-haematopoietic cells a protein similar to CD45 has been found; this has been termed LAR for leukocyte common antigen related. This protein has the same basic structure consisting of a large extracellular domain, a single transmembrane span and a cytoplasmic domain with homology to CD45 and protein tyrosine phosphatase 1B. Interestingly, the extracellular domain of LAR is related in sequence to the neural-cell adhesion molecule N-CAM. If this reflects a common behaviour of N-CAM and LAR, it is likely that LAR binds to cell surface molecules on neighbouring cells. These receptor tyrosine phosphatases would appear to constitute a new class of receptor.

2.5 Ligand gated ion channels

A final class of receptors whose signalling mechanisms are fairly well understood is the ligand gated ion channels. The best characterized of these is the nicotinic acetylcholine receptor. This receptor is a neurotransmitter dependent ion channel that is located on the post-synaptic side of nicotinic cholinergic synapses. Upon binding its ligand, the neurotransmitter acetylcholine, the receptor opens up an ion channel within its own structure. This channel allows the passage of sodium, potassium and calcium ions leading to a rapid depolarization of the membrane which provides an action potential to the post-synaptic cell.

The structure of the nicotinic acetylcholine receptor has been characterized in great detail, partly due to its ready availability in large quantities from the electric organs of electric eels and fish such as *Torpedo californica*. The receptor is a pentamer of four different subunits, α, β, γ and δ in the stoichiometry $\alpha_2\beta\gamma\delta$, with subunit sizes varying from 40,000 to 65,000. The primary structures of the subunits are related, with each having four highly hydrophobic stretches of amino acids plus a fifth amphipathic stretch which in an α-helix configuration would be hydrophobic on one side and hydrophilic on the other. In one model of acetylcholine receptor structure it is proposed that each subunit contributes five transmembrane α-helices to the receptor, making a total of twenty, arranged in such a way that the amphipathic helices form an aqueous pore down the middle to act as an ion channel. Low resolution quaternary structures of the nicotinic acetylcholine receptor have been obtained; they show that the five subunits are arranged as a torus with a central pore. The exact topology of the transmembrane spans has not, however, been resolved.

A number of other neurotransmitter receptors are ligand gated ion channels and some may be structurally related to the nicotinic acetylcholine receptor. In particular, the ligand binding subunit of the receptor for glycine, one of the major inhibitory neurotransmitters, shows considerable homology to the subunits of the nicotinic acetylcholine receptor with a similar arrangement of hydrophobic, putative membrane spanning regions.

3. Growth factor receptors with unknown signalling mechanisms

While the signalling pathways of a number of growth factor receptors are now at least partly understood, there are very many more receptors whose mechanism of action remains obscure, even after their primary structure has been determined. In this section consideration will be given to some of these enigmatic molecules.

3.1 Receptors for multifunctional growth factors

A number of peptide growth regulators exist that can have different effects on cell growth under different conditions. As was discussed in Chapter 2, transforming growth factor type β (TGFβ) falls into this category. It was initially identified by its property of promoting the growth of normally adherent cells in suspension but was subsequently found more often to inhibit the growth of responsive cells, particularly in monolayer culture. Despite intensive study, the receptors for TGFβ remain poorly characterized. Chemical cross-linking of ^{125}I-labelled TGFβ to cells reveals that there are three distinct receptors for the factor: one is a proteoglycan of molecular mass 280,000 while the other two are not heavily glycosylated and have molecular masses of 85,000 and 65,000 (see figure 7). The differing functions of these receptors is complicated by the fact that there are a number of closely related forms of TGFβ itself and more less closely related factors (see Chapter 2). The different peptides bind to the different receptors with varying affinities and in this way may transmit a spectrum of signals to the cell. An analogy can be drawn with insulin and the insulin-like growth factors.

While many cellular responses to TGFβ have been documented, very little is understood about the immediate signals passed by the receptors to cellular components. The mechanism of action of the TGFβ receptors will be particularly interesting because of the large number of growth and differentiation factors that are structurally related to TGFβ such as the mammalian inhibins, activins, Müllerian inhibitor and the decapentaplegic peptide in *Drosophila*.

Another group of peptide factors that can have either positive or negative effects on cell growth are the interferons. These molecules were originally identified by their antiviral effect although it is now clear that they have a much broader role. The family of interferons consists of the interferons type α and β (IFN-α-β), which bind to one receptor type of molecular mass 130,000, and interferon type γ (IFN-γ) which binds to a different receptor of molecular mass 90,000. This latter receptor has now been cloned: it shows no homology to any known receptor. The initial signalling mechanisms used by both types of interferon receptors remain obscure.

3.2 Immunological and haematopoietic receptors

In recent years the primary structures of a great many cell surface proteins found on haematopoietic cells have been determined. Most of these proteins are likely to be receptors of one sort or another. The ligands for these proteins may be soluble haematopoietic growth

Iodinate ligand TGF-β with ^{125}I

Allow labelled growth factor to bind to responsive cells in culture

Treat cells with a homobifunctional cross-linker

Lyse cells in detergent

Separate cell lysate on SDS polyacrylamide gel

Identify molecular weights of iodinated bands on autoradiograph

Subtract molecular weight of ligand to give
molecular weight of receptor

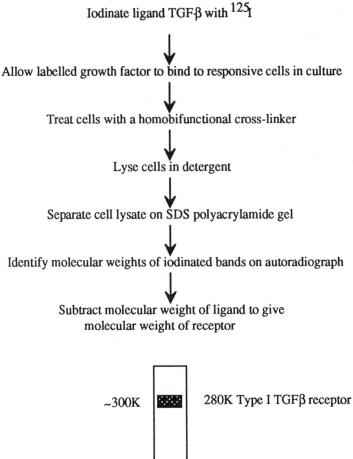

~300K   280K Type I TGFβ receptor

110K   85K Type II TGFβ receptor

90K   65K Type III TGFβ receptor

25K   uncrosslinked TGFβ

Fig 7. Identification of TGF-β receptors by cross-linking. The upper section shows the strategy and the lower section an idealised result indicating the iodine-labelled proteins.

factors such as the interleukins, soluble differentiation factors such as the colony stimulating factors (CSFs), cell surface molecules of other cells, (for example the major histocompatibility antigens or adhesion molecules), extracellular matrix proteins like fibronectin, or other soluble proteins such as immunoglobulins. With a few exceptions the mechanisms that these receptors use to transmit their diverse signals are not yet understood. It would be beyond the scope of this book to enter into a detailed discussion of the primary structures of all these molecules. Suffice it to say that in most cases these sequences have not been very informative in establishing functional mechanisms. The signalling mechanisms of a few well studied haematopoietic receptors are, however, reasonably well understood.

The antigen receptor on B-lymphocytes consists of surface immunoglobulin IgM or IgD molecules; the multiple usage of these molecules both as soluble antibodies and as cell surface receptors with an added transmembrane tail is an interesting example of the parsimonious reuse of specialized proteins. These cell surface immunoglobulins are coupled to the phosphatidyl inositol turnover pathway by a *pertussis toxin* insensitive G protein. It is not clear if this is a direct linkage or whether other molecules are involved, since the B-cell antigen receptor has no obvious similarity to the other receptors coupled to G proteins.

The antigen receptor of T-lymphocytes also appears to be coupled to phosphatidyl inositol turnover, but again it is possible that this is not direct. The T-cell receptor consists of up to 7 different subunits, two of which (α and β) are variable and the rest conserved (γ, δ, ε, η and ζ). There is no obvious structural similarity between the T-cell receptor and other receptors that are coupled to G proteins (see figure 3). Another aspect of the action of this receptor is that the intracellular tyrosine kinase p56^{lck} becomes activated upon antigen binding. As discussed in section 2.1 of this chapter, this kinase is associated with the cytoplasmic domain of the cell surface proteins CD4 and CD8. Interaction of the T-cell receptor (CD3) with CD4 or CD8 occurs during antigen binding.

3.3 Receptors as transport proteins

Within the cell there exist a great number of proteins that are involved in transporting macromolecules from one location to another. Many of these proteins are highly specific in what they bind and thus may be considered to be types of receptor proteins. Somewhat surprisingly a link between these types of transport proteins and cell surface receptors for growth factors has recently emerged. The receptor for the insulin–like growth factor type II (IGF-II) is identical to the cation-independent mannose–6–phosphate receptor; indeed mannose–6–phosphate binds to the IGF-II receptor and increases its affinity for IGF-II. The function of the mannose–6–phosphate receptor had previously been shown to be to target lysosomal enzymes from their site of synthesis to their final destination in the lysosome. This appears to involve recognition of particular carbohydrate structures on the enzymes, of which mannose–6–phosphate is a part. This unexpected identity of the two receptors has yet to be adequately rationalised. It is possible that the IGF-II receptor is not in fact involved in growth signal transduction but that the effects of IGF-II on cell growth are all mediated by the other receptors to which it binds such as the IGF-I receptor (see section 2.1 of this chapter). The IGF-II receptor may thus play a passive role in clearing excess IGF-II but importantly not insulin which does not bind to this receptor.

A number of other transport proteins have been suggested to have direct effects on cell growth beyond the simple supply of nutrients or other materials to the cell. The transferrin receptor which mediates the uptake of the iron carrying protein transferrin to the cell has been much scrutinized in this regard, but so far no clear evidence of a signal transduction pathway has emerged.

4. Oncogenes encoding receptor molecules

4.1 Oncogene products related to receptor tyrosine kinases

In Chapter 2 it was discussed how some oncogenes encode growth factors. This discovery went a great way towards explaining how activated oncogenes can subvert the normal growth control mechanisms of cells by causing them to produce their own growth factors in an autocrine manner, leading to unregulated proliferation. Soon after the discovery that the proto-oncogene c-sis encodes the growth factor PDGF, another important connection between an oncogene and a component of a growth signalling pathway was established when it was found that the sequence of peptides from the human receptor for the epidermal growth factor were highly homologous to the protein sequence of the product of the chicken oncogene v-erb B. It is now clear that the EGF receptor gene and the proto-oncogene c-erb B are one and the same. The significance of this is considerable since it establishes that growth factor receptors can be mutated in such a way as to make them capable of providing a continuous growth signal to a cell causing its transformation. The mutations by which the EGF receptor becomes constitutively activated in the avian erythroblastosis retrovirus are deletions at both the amino- and carboxy-terminal ends of the protein. The most important deletion is at the amino-terminal end where almost the whole of the extracellular ligand binding domain is removed (indeed the v-erb B encoded protein can no longer bind EGF). This deletion appears to lock the receptor into a conformation similar to the activated state even in the absence of ligand.

It is, in theory, possible that any receptor protein capable of generating a growth signal could be mutated in such a way as to make it constitutively activated. Alternatively, simple overexpression could achieve the same result. In practice, it is likely that only a very few mutations are capable of mimicking the conformational changes induced by ligand binding and attempts to artificially duplicate this process have been unconvincing. Nature has, however, provided us with other interesting examples of receptor tyrosine kinases acting as oncogene products.

Within the EGF receptor subfamily is the oncogene neu, also known as erb B-2. The transforming form of the neu protein, which was originally found by transfection of DNA from a rat neuroblastoma cell line, is constitutively activated by a different mechanism to that seen in v-erb B. Instead of a large truncation, the p185^{neu} has undergone a single point mutation in the transmembrane domain. One of the hydrophobic amino acids has been replaced by a charged residue. This mutation increases the tyrosine kinase activity of the protein, possibly by inducing its dimerization via the transmembrane domains. p185^{neu} has received considerable attention in recent years because of its apparent involvement in the genesis of human breast tumours where it is frequently found to be overexpressed.

Another well characterized example of a receptor that is encoded by a proto-oncogene is the receptor for the macrophage growth factor CSF-1. This protein was found to be closely related to the product of the feline retroviral oncogene *v-fms*. Subsequent studies have shown that the CSF-1 receptor, which is a member of the PDGF receptor subfamily, is indeed the product of the *c-fms* gene and that it can be activated to a transforming protein by the combination of a small deletion at the carboxy-terminus, which removes the site of an inhibitory tyrosine phosphorylation, and a point mutation in the extracellular domain. Unlike the *v-erb B* protein, the *v-fms* protein is still capable of binding its ligand but is, however, active whether or not CSF-1 is bound. It is evident that there are a number of different ways in which receptor tyrosine kinases can be mutated to form transforming proteins (see figure 8).

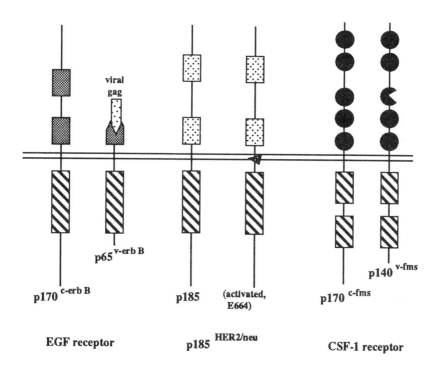

Fig 8. Models of activated forms of the receptor tyrosine kinases that are capable of causing cellular transformation. The hatched boxes represent the cytoplasmic tyrosine kinase domains, the grey boxes the cysteine rich extracellular domains and the black circles the immunoglobulin-like domains. The EGF receptor can be activated by truncation at both ends of the molecule, the neu/HER2 protein by point mutation in the transmembrane domain and the CSF-1 receptor by point mutation in the extracellular domain and C-terminal truncation.

Certain other proteins encoded by oncogenes found in retroviruses or discovered through transfection of DNA have been shown to encode proteins that would be expected to be receptor tyrosine kinases based upon their sequences. Ligands for these putative receptors have not yet been identified and in some cases little is known of the proteins themselves. These oncogenes include *kit, ret* and *sea,* which would be expected to encode members of the PDGF receptor subfamily, and *ros, trk* and *met* which may encode members of the insulin receptor subfamily.

4.2 Oncogene products related to G-protein coupled receptors

While the receptor tyrosine kinases have proved to be a very fertile source of oncoproteins, there is one other type of cell surface receptor that may have the potential to transform cells under certain circumstances, namely the G-protein coupled seven transmembrane span receptor family. An example of a protein of this type that has been implicated in cellular transformation is the product of the *mas* oncogene. This oncogene was isolated by transfection from a human epidermoid carcinoma. Later work showed that the activation of *mas* actually occurred as a result of the transfection procedure. Nevertheless *mas* is clearly capable of transforming fibroblasts albeit fairly weakly. The structure of the *mas* oncogene clearly showed that it was likely to be a receptor coupled to a G protein. Expression of the putative receptor in *Xenopus* oocytes resulted in the cells becoming responsive to peptides of the angiotensin family and the *mas* protein appeared to cause phosphatidyl inositol turnover in response to angiotensins. The *mas* oncogene thus appears to encode a receptor for angiotensin. It is not clear what mutation, if any, has led to its constitutive activation or if ectopic or excessive expression is responsible for its transforming ability.

Recent reports have also indicated that ectopic expression of the cloned $5HT_{1c}$ receptor can transform fibroblasts in the presence of serotonin. This protein is a typical G protein linked receptor that controls phosphatidyl inositol turnover.

5. A genetic approach: further insight into the role of receptors in growth and development

Attempts to study the function of receptors in vertebrate systems have been hampered by the weakness of genetic analysis in these organisms. By contrast, very powerful techniques of genetic analysis can be applied to many invertebrate and in particular unicellular eukaryotes. This includes the slime mould *Dictyostelium discoideum* which is a unicellular eukaryote that can, under certain conditions, aggregate to form a multicellular "slug" which contains two cell types. This organism has been intensely studied by genetic means as perhaps one of the simplest of all possible developmental systems. However the bulk of genetic analysis of development has been carried out on the fruit fly, *Drosophila*.

5.1 Receptor tyrosine kinases in *Drosophila*

A number of receptor tyrosine kinases have been identified in *Drosophila*. The first to be discovered was the homologue of the EGF receptor ("DER"). This protein is highly homologous to the human polypeptide with 41% overall identity. DER possesses all the typical characteristics of an EGF receptor subfamily member, also being similarly related to HER2/*neu*. Interestingly DER has three rather than two cysteine rich regions in the extracellular domain.

Two previously described recessive embryonic lethal phenotypes, *faint little ball* (*flb*) and *torpedo,* have now been shown to be mutations in the DER gene. The deletion of the EGF receptor gene leads to a complex embryonic phenotype in which there is a widespread failure of many structures to develop. The effects of smaller mutations in the DER gene are now being studied. It is hoped that second site reversion mutations will be discovered that will correct the phenotype due to defects in the EGF receptor by altering the function of other gene products; such mutations may reveal downstream targets of EGF receptor action.

The involvement of another receptor tyrosine kinase in *Drosophila* development is somewhat better understood. During development of the fly eye, cells form into clusters of eight photoreceptor cells plus twelve non-neuronal cells; these form the omatidia of the adult compound eye. A mutant was identified some years ago, named *sevenless,* in which one of these photoreceptor neurons, R7, fails to develop but instead becomes a cone cell. The fly develops normally in every other respect and even the effects on its vision are very subtle. Cloning and sequencing of the *sevenless* gene revealed that it encoded a possible receptor tyrosine kinase, most closely related to the *c-ros* proto-oncogene product. The *sevenless* protein is expressed in a number of different cell types in the eye including R7. The essential function that it appears to fulfil is to mediate the interaction of cell R7 with cell R8; this is necessary to induce R7 to develop into a photoreceptor. The protein on R8 that *sevenless* protein interacts with is either encoded by, or under the control of, the product of the *boss* (bride of sevenless!) gene, which was identified as having a similar phenotype to *sevenless.* Presumably this protein could activate the tyrosine kinase activity of the *sevenless* protein by binding to it as a ligand, causing a signal to be transmitted to the R7 cell determining its developmental fate. The distribution of the sevenless protein in the developing eye makes it very unlikely that it interacts with a soluble ligand but rather a cell surface protein on R8. If this is the case, this would be the first example of a growth factor receptor tyrosine kinase using a membrane protein as a ligand.

Other receptor tyrosine kinases studied in *Drosophila* include an insulin receptor homologue and also the product of the *torso* gene. This latter gene is involved in pattern formation in embryo development. The genetics again suggest that the *torso* protein is involved in cell–cell recognition.

5.2 G protein coupled receptors in yeast

While the ability to manipulate unicellular organisms such as yeast is very great, one might not expect to learn very much of relevance to human biology from such an evolutionarily distant life form. Fortunately this is in fact very far from the truth, since it appears that many basic cellular mechanisms are very similar between yeast and man. For example, the control of the cell cycle has recently been found to have a great many similarities (see Chapters 5 and 6). Yeast also possesses homologues of the *ras* oncogene family and has been shown to have tyrosine kinases.

The budding yeast, *Saccharomyces cerevisiae,* has been found to have two cell surface receptors that are very similar to the G protein linked receptors of mammalian cells. These are the receptors for the yeast peptide mating factors. The α-factor binds the product of the *STE2* gene while the a-factor binds to the product of the *STE3* gene. These genes, originally identified because mutations in them lead to a sterile phenotype, code for proteins with seven transmembrane spans and are clearly related to the adrenergic and muscarinic receptors in structure. Using the powerful genetic techniques available for yeast, a dozen genes have been identified that encode proteins that form part of the mating factor signal transduction pathway downstream of these receptors. These genes include three coding for the three subunits, α, β, γ, of a G protein that is similar to the heterotrimeric G proteins of mammalian cells. Both types of receptor interact with the same G protein, they are, however, expressed in different cells.

Major advantages of using the yeast system are the ease with which mutants can be generated and identified, the relative simplicity of isolating genes and the ability to "knock out" chosen genes. However, there can be problems in predicting just how relevant findings in yeast will be to the mammalian system. It appears to be quite common that parts of signalling pathways may be very similar between yeast and man while other parts of the same pathway may be quite dissimilar or perhaps twisted around in some way. For example, the G protein coupled to the mating type receptors does not function in the same way as most mammalian G proteins. It appears that the GDP bound α subunit binds to and inhibits the β subunit while the GTP bound α subunit releases the β subunit which itself then interacts with and activates its target. In mammalian systems the α subunit directly stimulates the effector, although in some special cases a system similar to the above has been proposed by some investigators (see Chapter 4). Although several mammalian signal transduction systems do not have parallels in yeast, for those that do, a great deal can be learnt from the exploitation of the genetic methods developed for use in this organism.

4

Second messengers

1. A large number of external signals regulate cellular processes through a small number of second messengers

The previous chapter described how external signals bind to their specific receptors. The consequence of this binding can be a change in the enzyme activity contained within the receptor polypeptide or a separate molecule. Some receptor types cause changes in the activity of enzymes that catalyse the production or degradation of second messengers. Second messengers are small intracellular molecules or ions that can transmit a signal within the cell. While an external signal (in this conception: the first messenger) carries the information from the environment to the cell surface, a second messenger provides a link with proteins that regulate intracellular processes. So far, only a few second messengers have been discovered, nevertheless, these second messengers transmit information from a large number of extracellular signals [Table 1]. Thus, they are a common component of many different regulatory pathways.

The role of second messenger has been attributed to 3',5'-cyclic adenosine monophosphate, 3',5'-cyclic guanosine monophosphate, 1,4,5-inositol trisphosphate, diacylglycerol and Ca^{2+}. The structure of these molecules is presented in figure 1.

Cyclic adenosine monophosphate (cAMP) is produced from the precursor molecule adenosine triphosphate (ATP)(figure 2a). The reaction is catalysed by the enzyme adenylyl cyclase. External signals can stimulate or inhibit this enzyme and thus control the production of cAMP. Another molecule that acts as a second messenger is cyclic guanosine monophosphate (cGMP) (figure 2c). The enzymes that catalyse production (guanylyl cyclase) and degradation (cGMP phosphodiesterase) of cGMP can both be regulated by external signals.

Hydrolysis of a membrane phospholipid phosphatidyl inositol 4,5-bisphosphate (PIP_2), generates two second messengers: 1,4,5-inositol trisphosphate (IP_3) and diacylglycerol (DG)(figure 2b). This reaction is catalysed by a phospholipase C. Many external signals can increase the hydrolysis of PIP_2 by stimulating phospholipase C activity.

Another enzyme involved in hydrolysis of cellular phospholipids, phospholipase A_2

a)

CH₃
CH₂
CH₃   CH₂
CH₂   CH₂
CH₂   CH₂
CH₂   CH
CH₂   CH
CH₂   CH₂
CH₂   CH
CH₂   CH
CH₂   CH₂
CH₂   CH
CH₂   CH
CH₂   CH₂
CH₂   CH
CH₂   CH
CH₂   CH₂
CH₂   CH₂
CH₂   CH₂
C=O   C=O
O   O
C — C — C
              OH

c)

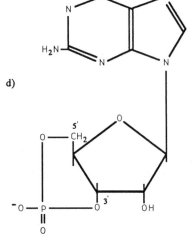

b)

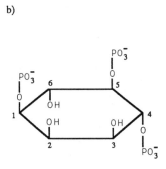

d)

Fig 1. The structure of some second messenger molecules: a) diacylglycerol; b) 1,4,5-inositol trisphosphate; c) 3',5'-cyclic adenosine monophosphate d) 3',5'-cyclic guanosine monophosphate.

Table 1. Selected examples of second messenger mediated responses

a) Responses mediated by cAMP

External signal	Cell type	Up/down	Response
adrenaline (β-receptor)	skeletal muscle	up	breakdown of glycogen
adrenaline (β-receptor)	fat cells	up	breakdown of lipids
thyroid stim. hormone	thyroid gland	up	secretion of thyroxine
glucagon	liver	up	breakdown of glycogen
serotonin	salivary gland	up	fluid secretion
odorant	olfactory epithelium	up	olfaction
tastant	lingual epithelium	up	gustation
adrenaline (α$_2$-receptor)	blood platelets	down	aggregation and secretion
adrenaline (α$_2$-receptor)	fat cells	down	decrease of lipid breakdown
adenosine	fat cells	down	decrease of lipid breakdown
nutritional signal	*S. cervisiae*	up	proliferation
adrenaline (β-receptor)	parotid cells	up	proliferation
adrenaline (β-receptor)	liver (regeneration)	up	proliferation
thyroid stim. hormone	thyroid gland	up	proliferation
growth hormone RF	pituitary somatotrophs	up	proliferation

b) Responses mediated by cGMP

External signal	Cell type	Up/down	Response
resact or speract	sea urchin spermatozoa	up	chemotaxis
atrial natriuretic pep.	vascular smooth muscle	up	relaxation
light	retinal rod cells	down	vision

c) Responses mediated by IP$_3$/DG

External signal	Cell type	Up/down	Response
acetylcholine	pancreas	up	amylase secretion
acetylcholine	pancreatic β-cells	up	insulin secretion
acetylcholine	smooth muscle	up	contraction
thrombin	blood platelets	up	aggregation
fMet-Leu-Phe	neutrophils	up	chemotaxis
antigen	mast cells	up	histamine secretion
vasopressin	liver	up	breakdown of glycogen
light	*Drosophila* photoreceptor	up	vision
spermatozoa	sea urchin eggs	up	fertilization
bombesin	fibroblasts	up	proliferation
thrombin	fibroblasts	up	proliferation
angotensin II	fibroblasts	up	proliferation
bradykinin	fibroblasts	up	proliferation
PDGF	fibroblasts	up	proliferation
EGF	fibroblasts	up	proliferation

(PLA$_2$), can also be the subject of regulation by external signals. PLA$_2$ catalyses hydrolysis of the C-2 ester bond of phospholipids to produce fatty acid (e.g. arachidonic acid) and the lysophosphatidyl derivative of the phospholipid. Arachidonic acid itself has not been shown to function as a second messenger, however it can be further metabolized to biologically active compounds such as prostaglandins, thromboxanes, leukotrienes and lipoxins.

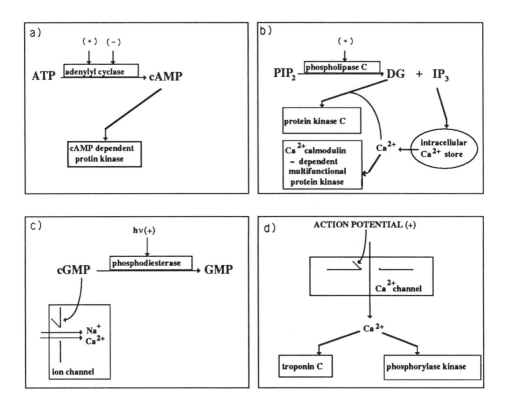

Fig 2. Second messengers: production and targets. Upper figures illustrate the production of (a) cAMP and of (b) DG and IP$_3$. In a large number of cell types, the intracellular targets for these second messengers are cAMP-dependent protein kinase, protein kinase C and through the mobilization of Ca^{2+}, multifunctional Ca^{2+}/calmodulin-dependent protein kinase. However, some specialized cells have specific regulatory pathways (lower figures). In retinal rod cells (c) cGMP binds to and opens an ion channel. Reduction of cGMP concentration after stimulation by light causes closure of the channels. The resulting hyperpolarization of the membrane provides a neuronal signal. In skeletal muscle (d), nerve impulses trigger an action potential at the plasma membrane that causes opening of Ca^{2+} channels. Ca^{2+} diffuses into the sarcoplasm and interacts with a number of Ca^{2+} binding proteins including, troponin C and calmodulin. In this case some calmodulin is already tightly associated as a subunit of phosphorylase kinase. Binding of Ca^{2+} to troponin C causes the muscle to contract; the Ca^{2+}/calmodulin interaction causes activation of phosphorylase kinase with the consequent activation of phosphorylase and glycogen breakdown. Thus, the energy requirement for muscle contraction can be replenished through glycogenolysis, the two being coordinately regulated by Ca^{2+}.

Changes in the concentration of intracellular Ca^{2+} have been recognized as an early event in the action of external signals. The concentration of Ca^{2+} in the cytoplasm can be increased by entry of the ion from the extracellular environment (figure 2d). Many hormones and neurotransmitters, however, can increase the concentration of Ca^{2+} in the cytoplasm by mobilizing the ion from intracellular stores. This mobilization is induced by one of the products of PIP_2 hydrolysis, IP_3 (figure 2b).

These second messenger molecules bind to specific intracellular proteins (figure 2). A number of these target proteins possess protein kinase activity or are directly involved in the regulation of such an activity. Cyclic nucleotides, cAMP and cGMP bind to and activate protein kinases named accordingly as cAMP-dependent and cGMP-dependent protein kinase. The second messenger DG activates a protein kinase named protein kinase C. Many actions of Ca^{2+} are mediated through the formation of a complex with the protein calmodulin. This complex can regulate the activity of several molecules, including protein kinases (e.g. multifunctional Ca^{2+}/calmodulin-dependent protein kinase, myosin light chain kinase and phosphorylase kinase). In some cases, the targets for second messengers have other functions. For example, cGMP in retinal rod cells binds to and regulates the activity of a specific type of ion channel (figure 2c); the Ca^{2+} released in skeletal muscle binds to troponin C, the subunit of troponin responsible for regulation of muscle contraction (figure 2d).

A detailed discussion of the major target molecules for second messengers, the protein kinases, will be presented in Chapter 5. In the following sections of this chapter, we shall describe transmembrane reactions that transduce an extracellular signal into a change of second messenger concentration. We shall also describe how second messengers participate in the regulation of cell growth. Finally, we shall focus on the molecules participating in these growth regulation pathways that are altered in transformed cells.

2. Receptors can transduce an external signal into a change in the activity of an effector via the action of a GTP-binding protein

Studies on the mechanism(s) of signal transduction have been of major interest since the discovery of cAMP and its role as a second messenger in hormone mediated responses. An important observation was made when isolated plasma membranes were used to study this phenomenon instead of intact cells. Binding of the hormone to its receptor was not by itself sufficient for the regulation of adenylyl cyclase in the membranes. Another ligand, from inside the cell, was required. This ligand was identified as GTP. The binding site for GTP was neither on the receptor nor on the effector enzyme, adenylyl cyclase, but on a separate molecule. This third component of signal transduction was designated G-protein (G = GTP-binding protein). Studies on the stimulation of cGMP phosphodiesterase in retinal rod cells resulted in similar findings. Detailed analysis of those two systems revealed several features that appeared to be specific for a G-protein mediated signal transduction. For example, non-hydrolysable analogues of GTP and toxins of *Vibrio cholerae* and *Bordatella pertussis* interfere with both of these responses (in figure 3 their effects are discussed in the context of a model for signal transduction). These observations, together with the essential requirement for GTP, are of important experimental value for identifying new processes mediated by G-proteins. For example, G-proteins have also been implicated in the control of production of the second messengers IP_3 and DG. Furthermore, the enzymes regulating levels of second messengers are not the only type of effector molecules. Ion channels involved in the change of the membrane potential (itself a potent regulator of cellular function) can be coupled to receptors via G-proteins. It is currently estimated that about 40 different external signals that bind to about 100 different receptors[1], employ a G-protein mediated mechanism of signal transduction. This mechanism of signal transduction is widely distributed among eukaryotes, from the unicellular yeasts to mammals.

The molecular details of the interaction between receptor, G-protein and effector, have been derived mainly from two experimental systems. These are adenylyl cyclase stimulated by hormone receptors and cGMP phosphodiesterase stimulated by rhodopsin (light). A model based on these two systems appears to be applicable to a number of related signal transduction pathways. According to this model (figure 3) the participating molecules transmit information by inducing an alteration in the conformation, and consequently the function, of the protein next in line. Binding of the extracellular signal to the receptor induces a conformational change that promotes interaction of the receptor with a specific G-protein. This interaction leads to a conformational change of the G-protein that permits exchange of GDP for GTP. The GTP-bound G-protein undergoes another conformational change that enables it to interact with the appropriate effector molecule, and so modify its activity.

During this process of signal transduction the initial signal is amplified. At the first step of amplification, one activated receptor interacts with and activates several G-protein molecules. The second step of amplification occurs during the lifetime of a G-protein/effector complex. An activated enzyme can produce (or destroy in the case of cGMP phosphodiesterase) several hundred second messenger molecules.

Table 2 summarizes specific features of G-protein mediated signal transduction controlling: (A) the concentration of cGMP in retinal rod cells, (B) the concentration of cAMP and (C) the concentration of IP_3 and DG in many different cell types. The properties of some components participating in regulation of ion channels are also included (D).

A more detailed discussion of the mechanism of signal transduction requires knowledge of the structure and function of the participating proteins. In the following text we shall present the properties of receptors, G-proteins and effectors controlling the level of second messengers and describe what is currently known of their interaction.

---

[1] The reason that the number of distinct receptors exceeds the number of external signals is the existence of subtypes. This is well exemplified by at least four subtypes of adrenergic receptors (α_1, α_2, β_1, and β_2). For details see Chapter 3.

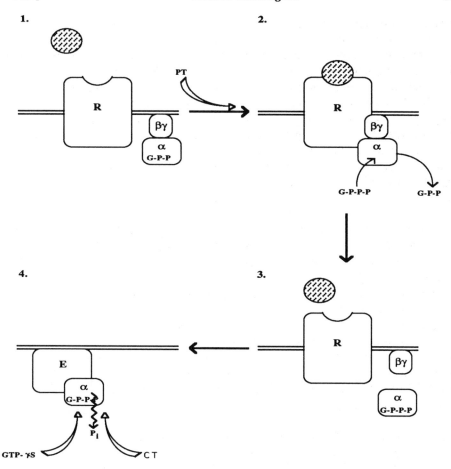

Fig 3. Interactions between receptor, G-protein and effector.

1. Receptor (R) and G-protein (α and βγ subunits) are shown in the basal (resting) state. Subunits of the G-protein are tightly associated, and the α subunit is in its inactive, GDP-bound state.

2. Binding of the hormone activates the receptor which can now interact with a G-protein. This receptor/ G-protein interaction allows a rapid exchange of GTP for GDP in the α subunit.

3. The binding of GTP to the hormone-receptor-G-protein complex has two major consequences. Firstly, it causes dissociation of the receptor from both G-protein and hormone. Secondly, binding of GTP markedly decreases affinity of the α subunit for the βγ complex.

4. Activated α-GTP can regulate the activity of an effector molecule (E). The change in the activity of the effector molecule is only transient because the GTP-ase activity inherent in the α subunit hydrolyses GTP to GDP, converting it back to the inactive conformation that is no longer able to interact productively with the effector; α-GDP reassociates with βγ and enters another cycle.

White arrows indicate sites at which this normal cycle of signal transduction can be interrupted. *Cholera toxin* (CT) inhibits GTP-ase activity of some α subunits and so prolongs the half-life of the GTP-bound state. As a result the effector responds continually, even in the absence of external signal. A similar effect is produced by the addition of analogues of GTP (e.g. GTP-γ-S) which bind to α subunits instead of GTP but can not be hydrolysed. *Pertussis toxin* (PT) can inhibit signal transduction by blocking interaction between certain G-proteins and their receptors.

Table 2. Specific features of some molecules participating in G-protein mediated signal
   transduction

A) Regulation of cGMP concentration in retinal rod cells

Molecules mediating transmission of the visual signal in retinal rods have been extensively
studied. Although specific for this cell type, they exemplify recurring structural motifs typical
of G-protein mediated signal transduction.

Components:

Receptor. Rhodopsin with the chromophore 11-*cis*-retinal. Integral membrane protein (40
kDa). Structure typical for G-protein coupled receptor (i.e. seven transmembrane α-helices).
G-protein. Transducin, Gt. Multisubunit peripheral membrane protein: α_t (39 kDa), β (36 kDa),
and γ (~10 kDa).
Effector. cGMP phosphodiesterase. Multisubunit peripheral membrane protein: α (88 kDa),
β (85 kDa), and γ (9 kDa).

B) Regulation of cAMP concentration

Regulation of cAMP concentration can be achieved through stimulation or inhibition of
adenylyl cyclase. Changes in cAMP concentration mediate the action of many external signals
in different cell types, including those that regulate cell growth. The mechanism of stimulation
of adenylyl cyclase and the components that participate in the process are understood better
than those of the inhibitory pathway.

Components:

Stimulatory receptor. Several, e.g. β_1- and β_2-adrenergic. Integral membrane glycoproteins.
Structure typical for G-protein coupled receptors (i.e. seven transmembrane α-helices).
Stimulatory G-protein. Gs. Multisubunit peripheral membrane proteins: α_s (45 or 52 kDa), β
(35 or 36 kDa), and γ (~10 kDa).

Inhibitory receptor. Several, e.g. α_2-adrenergic, muscarinic M2. Integral membrane
glycoproteins. Structure typical for G-protein coupled receptors (i.e. seven transmembrane α-
helices).
Inhibitory G-protein. G-protein(s) is substrate for pertussis toxin (G_i forms and G_o have been
implicated). Multisubunit peripheral membrane proteins.

Effector. Adenylyl cyclase. Integral membrane glycoprotein (120 kDa).

C) Production of IP$_3$ and DG

Second messenger roles for IP$_3$ and DG have been discovered relatively recently. The components involved in regulation of IP$_3$ and DG concentration have been studied in different systems (including regulation of cell growth). However, molecular properties of some components and the precise roles they employ in the signal transduction remains unclear. Here is described a G-protein mediated mechanism that may control production of IP$_3$ and DG by some (but not all) external signals.

Components:

Receptor. Many. Those structurally defined include muscarinic receptors (M1 and M4), substance K receptor, and 5-hydroxytryptamine receptor (1c). Structure typical for G-protein coupled receptor (i.e. seven transmembrane α-helices).
G-protein. Perhaps more than one. Involvement of GTP-binding proteins other than G-proteins can not be excluded. In some cases, PLC linked G-proteins are pertussis toxin substrates pointing to G_{i-1}, G_{i-2}, G_{i-3} and G_o as candidates.
Effector. Phospholipase C specific for inositol phospholipids. Several forms of the enzyme have been structurally defined. All are peripheral membrane proteins (60-150 kDa). At least one form (β-like PLC) can be regulated by G-proteins (see text).

D) Regulation of ion channels

Regulation of the activity of some types of K$^+$ and Ca^{2+} channels can be achieved through G-protein mediated pathways. One example described here is regulation of voltage-dependent, dihydropyridine-sensitive Ca^{2+} channel in cardiac and skeletal myocytes. Stimulation of this channel results in the increase of Ca^{2+} permeability and elevation of cytoplasmic Ca^{2+} concentration. In addition to stimulation by the change in membrane potential (that is independent of G-proteins), the channel can also be stimulated by hormones. The β-adrenergic stimulation employs a G-protein mediated mechanism of signal transduction.

Components:

Receptor. β-adrenergic. Integral membrane glycoproteins. Structure typical for G-protein coupled receptor (i.e. seven transmembrane α-helices).
G-protein. Gs. Multisubunit peripheral membrane proteins: α_s (45 or 52 kDa), β (35 or 36 kDa), and γ (~10 kDa).
Effector. Voltage-dependent, dihydropyridine-sensitive Ca^{2+} channel. Complex subunit structure (α, α_1, β, γ, δ). The α subunit, responsible for Ca^{2+} current, is an integral membrane glycoprotein (~200 kDa). Four large hydrophobic regions (each containing six putative transmembrane α-helices) of the α subunit are assumed to surround the ionic channel.

2.1 Receptors for different signals share common structural features

The G-protein coupled receptors belong to a number of different pharmacological classes. The ligands they bind are as divergent as biogenic amines, peptide hormones, are the chromophore retinal. In addition to pharmacological characterization, molecular properties of many G-protein coupled receptors are now known.

Rhodopsin was the first G-protein coupled receptor to be well characterized structurally. Two other receptor types that have been extensively studied are the adrenergic receptors (which mediate the physiological effects of catecholamines) and the muscarinic acetylcholine receptors. The β_1- and β_2-adrenergic receptor subtypes stimulate the activity of adenylyl cyclase. Conversely, the α_2-adrenergic receptors inhibit this enzyme. The α_1-adrenergic receptors stimulate hydrolysis of phosphoinositides by activating a specific phospholipase C. The subtypes of muscarinic acetycholine receptors are also involved in the regulation of different effector molecules mediating inhibition of adenylyl cyclase, stimulation of phospholipase C and the opening of K^+ channels. Other G-protein coupled receptors with defined primary structures include receptors for the biogenic amine 5-hydroxytryptamine, receptors for neural peptides (e.g. substance K) and receptors for mating peptides isolated from yeast cells.

Comparison of the amino acid sequences of G-protein coupled receptors indicates that receptors for different agonists have some structural features in common. The most consistently conserved of these features is the existence of seven clusters of hydrophobic amino acids. It is proposed that these hydrophobic segments form α-helices long enough to span a lipid bilayer. Therefore, a general topology for those receptors (i. e. their arrangement with respect to the membrane) can be described as seven membrane spanning domains connected by extracellular and intracellular loops[2]. The transmembrane topology of G-protein coupled receptors, exemplified by α_2- and β_2- adrenergic receptors, is illustrated in figure 4.

Further studies of the receptor molecules are directed towards characterization of functional domains within the general receptor structure. Especially important are those domains involved in ligand binding and interaction with G-proteins. Studies of the functional domains combine biochemical methods with use of immunological tools and the manipulation of protein structure via changes in the DNA encoding the protein. They have revealed the relationship between some specific residues and segments of the receptors with the receptor functions. Thus, the ligand binding site appears to be situated within a pocket in the plasma membrane formed by all seven hydrophobic α-helices. In the case of β-adrenergic receptors, the residues that determine specificity for a particular ligand are contained within the seventh membrane spanning domain. In this case, interaction with G-protein requires the third intracellular loop of the receptor (figure 4).

---

[2] It should be noted that this is not an arbitrary prediction of topology since low resolution structural information has been obtained for bacteriorhodopsin which like the mammalian protein is a seven transmembrane domian protein (see Chapter 3).

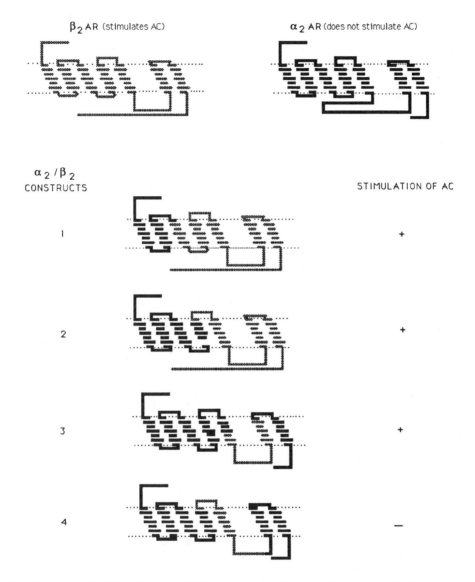

Fig 4. Genetic manipulation of receptor DNA: an approach to functional domain analysis. The construction of chimeric molecules of β_2- and α_2-adrenergic receptors (β_2AR and α_2AR) has contributed towards an understanding of the structural basis for the interaction of receptors with G-proteins. In this series of experiments, DNA encoding β_2AR and α_2AR and the DNA constructs combining the two (1-4), were used to make mRNA which was then injected into frog oocytes. Receptor proteins synthesized in the oocytes became inserted into the plasma membrane and thus could be stimulated with various agonists. Binding of adrenaline to β_2AR resulted in stimulation of adenylyl cyclase mediated by the endogenous (oocyte) G_s. The replacement of different portions of the β_2AR with equivalent portions of the α_2AR (constructs 1-4) and subsequent measurement of adenylyl cyclase stimulation was used to determine the part of the β_2AR responsible for coupling with G_s. The smallest segment of β_2AR that still enabled interaction of a receptor construct with G_s consisted of parts from the fifth and sixth membrane spanning domains and the third intracellular loop (construct 3). Any of the tested replacements within this segment of β_2AR diminished stimulation of adenylyl cyclase (e.g. construct 4). (According to *Koblika et. al. /1988/ Science 240 : 1310*).

2.2 The G-protein family: coupling of receptor to effector molecules

A major group of proteins that can fulfill the function of coupling receptors to effector molecules are G-proteins. Members of the family and their specific and common features are discussed below. Also described are some distantly related GTP-binding proteins that have been implicated in signal transduction.

Members of the G-protein family

Studies of the structural and functional properties of individual G-proteins began with their purification. Among the first to be purified and characterized was G_s, a stimulatory regulator of adenylyl cyclase. As described in figure 5, purification was achieved by following an activity able to reconstitute signal transduction in membranes lacking G_s. Some other G-proteins have been purified according to their ability to bind and hydrolyse guanine nucleotides or as substrates for bacterial toxins, leaving open the question of their biological role.

Structural analysis of the purified G-proteins has shown that they are heterotrimers composed of subunits designated α (39,000-52,000), β (~35,000), and γ (~10,000) according to decreasing molecular mass (figure 5). The α polypeptides clearly differ among G proteins and therefore contribute to the specificity of each G-protein (see below).

Determination of the amino acid sequence of several purified G-proteins revealed regions of diversity as well as regions of identity. Oligodeoxynucleotides corresponding to the conserved regions have proved useful in screening cDNA libraries for clones encoding related G-proteins. As a result, the family of G-proteins has grown larger than first anticipated (Table 3).

There are four forms of the α polypeptide of G_s (α_s) produced by alternative splicing of mRNA encoded by a single gene. The functions of these different α_s have been tested in reconstituted systems by adding each of the purified forms to potential effector molecules and measuring changes in the effector activity. The experiments have shown that all forms can not only stimulate adenylyl cyclase in the system originally used for purification of G_s, but also are able to activate a specific type of Ca^{2+} channel. This clearly demonstrates that one G-protein (in this instance any of the G_s forms) has the potential to regulate more than one effector.

G_{olf} shares considerable similarity (88% of identical amino acid residues) with G_s. In reconstituted systems, this G protein can stimulate adenylyl cyclase. G_{olf} is exclusively expressed in sensory neurons of olfactory neuroepithelium and it is therefore likely that the function of G_{olf} is to mediate stimulation of adenylyl cyclase in response to odorants (G_{olf} = olfaction).

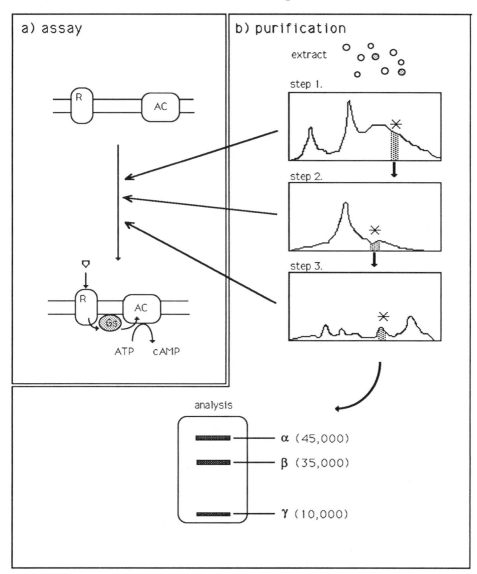

Fig 5. Purification of G$_s$.

a) G-proteins that stimulate adenylyl cyclase are best detected by an assay system deficient in G-protein. Such an assay is provided by the membranes of the cyc⁻ S49 lymphoma cell mutant. The mutant is deficient in G$_s$ activity but contains the enzyme adenylyl cyclase; due to the absence of G$_s$, adenylyl cyclase is in its inactive state. Addition of preparations containing functionally intact G$_s$ to cyc⁻ membranes fully restores adenylyl cyclase activity stimulated by hormone and guanine nucleotide.

b) Separation of G$_s$ from other proteins present in membrane preparations of the wild type, hormone responsive cells was achieved by combining several chromatographic steps. Chromatograms show the distribution of the protein in the fractions (solid line). Stars indicate those fractions which can stimulate adenylyl cyclase in cyc⁻ membranes and therefore contain G$_s$. Fractions with G$_s$ activity were used as starting material for the next purification step. After the third step of the preparation of G$_s$, analysis by SDS gel electrophoresis (the method separates individual polypeptides according to their size) revealed that it consists of three different polypeptides (see text).

| G-protein | properties of α subunit | |
name*	mol. mass x10^{-3}	modification by bacterial toxin
G$_s$ (α$_s$βγ) four α$_s$ polypeptides	45, 52	Cholera toxin
G$_{olf}$ (α$_{olf}$βγ)	45	Cholera toxin
G$_{i-1}$ (α$_{i-1}$βγ) G$_{i-2}$ (α$_{i-2}$βγ) G$_{i-3}$ (α$_{i-3}$βγ)	41 40 41	Pertussis toxin Pertussis toxin Pertussis toxin
G$_o$ (α$_o$βγ)	39	Pertussis toxin
G$_z$ (α$_z$βγ)	41	?
G$_{tr}$ (α$_{tr}$βγ) G$_{tc}$ (α$_{tc}$βγ)	41 40	Cholera toxin, Pertussis toxin Cholera toxin, Pertussis toxin

Table 3. Some of the mammalian G-protein molecules.
* The nomenclature for G-proteins was originally chosen to reflect their presumed biological role. The discovery of new members and new functions for the members previously assigned to a specific signal transduction system (and named accordingly) made this nomenclature inadequate but has yet to be changed. Functions of each of these structurally defined proteins are discussed in the text.

There are three highly homologous G$_i$ α polypeptides encoded by different genes, which define three G$_i$ proteins, namely G$_{i-1}$, G$_{i-2}$ and G$_{i-3}$. A related protein G$_o$ (o = other G protein), shares many properties with G$_i$, including modification by *pertussis toxin*. However, sequence homology between α$_o$ and α$_i$ forms is not as great as that shared by α$_{i-1}$, α$_{i-2}$, and α$_{i-3}$. The G-proteins modified by *pertussis toxin* have been implicated in the inhibition of adenylyl cyclase (G$_i$ = inhibitory regulator of adenylyl cyclase). However, their function is likely to extend to the regulation of other processes known to be affected by this toxin, such as gating of K$^+$ and Ca^{2+} channels and perhaps stimulation of phospholipase C. More direct evidence for the involvement of this group of G-proteins in the regulation of ion channels has been provided by reconstitution experiments. For example, addition of purified α$_{i-3}$ to membrane patches containing K$^+$ channels can cause their opening.

There are also two highly homologous forms of G$_t$ (t = transducer of the visual signal). It has been demonstrated, using selective antibodies which specifically recognize these proteins, that one form exists exclusively in retinal rods (G_{tr}) and the other only in cones (G_{tc}). Retinal rod cells contain receptors for black-white vision (rhodopsin) while retinal cone cells contain receptors for colour vision (colour opsins). Reconstitution experiments have shown that G_{tr} can couple photoactivated rhodopsin to cGMP phosphodiesterase. G_{tc} is presumed to have the same function (i.e. the function of coupling colour opsins to cGMP phosphodiesterase). It is therefore likely that G_{tr} and G_{tc} represent cell-type specific isoforms of the same protein.

In addition to the mammalian G-proteins discussed above, related G-proteins have been isolated from other organisms including the yeast *Saccharomyces cerevisiae* and the slime mould *Dictyostelium discoideum*.

Subunits of G-proteins

There are about 10 different types of α subunit. It is this subunit of G-proteins which goes through guanine nucleotide dependent cycles of activation and inactivation (figure 3). This process is common to all α subunits and also for many otherwise unrelated proteins. These proteins are collectively designated as GTP-binding proteins (see below). The structural basis for the binding of guanine nucleotides, the GTP-ase activity, and the conformational changes have been studied extensively in all members of the group. These studies have employed a variety of methods including structural analysis of deficient phenotypes (mutation mapping), X-ray crystallography, and manipulation of protein structure via changes in the nucleotide sequence of the cDNA (figures 6 and 9). As a result it is possible to identify some regions and individual amino acid residues within the polypeptide chains of this group of proteins, participating in guanine nucleotide binding and hydrolysis of GTP. In figure 6a, four such regions are highlighted in the primary structure of an α polypeptide. Some residues within these regions are conserved among all GTP-binding proteins studied and lie in the vicinity of the guanine nucleotide binding site.

The α subunits of G-proteins hydrolyse GTP at an apparently invariable rate, independent of interaction with other proteins. This feature distinguishes G-proteins from some other GTP-binding proteins (e.g. $p21^{ras}$ that interacts with GAP, see below). It has been suggested that the structural basis for this high intrinsic rate of GTP hydrolysis may be contained within the largest region separating the conserved (GTP binding) residues.

A common characteristic of many G-proteins is NAD-dependent ADP-ribosylation catalysed by *cholera* and/or *pertussis toxin*. The effects of this covalent modification on the function of G-proteins have been discussed earlier (see figure 3). The sites modified are located on the α subunit. Subunits α_s and α_t are modified by *cholera toxin* on an arginine residue; subunits α_i, α_o and α_t are ADP-ribosylated by *pertussis toxin* on a cysteine residue near the carboxy-terminus.

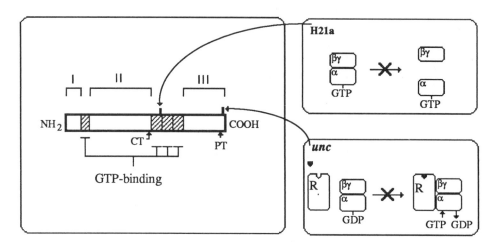

Fig 6. Defects in the function of α_s: mapping mutations in order to locate functional regions. On the left, a schematic outline of the structure of an "average" α subunit shows regions with highly conserved sequences (hatched boxes) and regions (I, II and III) where sequences are variable between different α subunits. Within conserved sequences are residues shared with the GTP-binding proteins involved in protein synthesis (e.g. EF-Tu) and the product of the *ras* oncogene (p21 *ras*). That these sequences play important roles in the structure of the guanine nucleotide binding site has been confirmed by x-ray crystallography of EF-Tu and p21*ras*. Also shown are the sites modified by *cholera toxin* (CT) in α_s and α_t and by *pertussis toxin* (PT) in α_i, α_o and α_t.
One of the methods which has contributed towards an understanding of the structural basis for the functions of α subunit has been the mapping of mutations in defective phenotypes of α_s. On the right it is shown that the α_s from the mutant cell line designated H21a can not activate adenylyl cyclase after being stimulated to exchange GDP for GTP. In this mutant the binding of GTP does not result in conformational change of α_s and dissociation of $\beta\gamma$. Mutation H21a (replacement of a glycine for an alanine) has been located within one of 4 regions of homology among different GTP-binding proteins. The location of this mutation suggests that the glycine residue involved may serve as a hinge that allows relative movement of separate domains of the protein. The glycine to alanine substitution in the mutant would alter flexibility of the hinge and prevent the conformational change.
A second mutant phenotype of α_s, *unc*, does not respond to stimulation by the β-adrenergic receptor. A similar effect is created by *pertussis toxin* catalysed ADP-ribosylation of α_i, α_o, and α_t. Both the *unc* mutation in α_s (replacement of a proline with an arginine) and the site of *pertussis toxin* catalysed ADP-ribosylation in other α subunits are located at the carboxy-terminal region of the α polypeptides. This, together with other lines of experimental evidence, points to the carboxy-terminal region as an interaction site of G-proteins with receptors. (Based on *Miller et. al. /1988/ Nature 334: 712 and Sullivan et. al. / 1987/ Nature/330:758*).

Another covalent modification of at least some α subunits is attachment of N-linked myristic acid at the amino terminus. The α subunit is not a hydrophobic protein and does not contain within its structure potential transmembrane domains. Thus, fatty acid acylation (myristic acid) can facilitate interaction between α subunits and the lipid bilayer as observed for the *src* family of protein tyrosine kinases (Chapter 5).

In addition to these common functions, each α subunit interacts with a set of specific receptor and effector molecules. Comparison of the primary structures of different α polypeptides reveals a consistent pattern of divergent sequences separated by conserved regions. Regions I, II and III, shown in figure 6 (left panel), contain stretches of greatest sequence variation which could provide a structural basis for the specific interaction of each α subunit. Indeed, the divergent carboxy-terminal stretch (within region III) has been identified as a site involved in the interaction with a receptor; one of the approaches that has provided the experimental evidence for this is illustrated in figure 6.

The subunits β and γ are, under physiological conditions, tightly but non-covalently associated. While α subunits are unique for each G-protein, β and γ appear to be structurally more similar and could be shared by at least some members of the family. In reconstitution experiments the β/γ complexes derived from G_s, G_i and G_o were functionally interchangeable. Only two forms of β polypeptide were originally identified; these are closely related (90% identical amino acid residues) and each form can be associated with any α subunit. A third β subunit has recently been identified. There is also evidence for the existence of more than one form of γ subunit. There are several functions of G-proteins that involve the β/γ complex. The hydrophobic nature of the complex can facilitate an interaction between a G-protein and the plasma membrane. When associated with the α subunit, the β/γ complex participates in the interaction with activated receptor; neither α nor β/γ alone can do so. After disso- ciation from activated α-GTP subunit, the β/γ complex can enter another cycle of signal transduction as part of a reassembled G-protein. Until then, as suggested by recent experimental evidence, free β/γ complex could cause a change in activity of some effector molecules. One form of phospholipase A_2, which catalyses the hy- drolysis of membrane phospholipids to produce arachidonic acid, may be regulated by β/γ rather than by an α subunit. As described earlier, arachidonic acid is a pre- cursor of several important, biologically active compounds.

Related GTP-binding proteins

The G-proteins are only one of several different families of GTP-binding proteins. Other such families include tubulins, elongation and initiation factors of protein synthesis and a group of proteins designated as "small GTP-binding proteins" exemplified by the products of *ras*, *rho*, *ral*, SEC4, and *arf* genes. These proteins have different structural properties, their molecular mass varying from 21,000 (for p21^{ras}) to 90,000 (for some elongation factors). As discussed earlier, they share a general function of GTP-binding and hydrolysis which controls the change between two different protein conformations. These GTP-binding protein families participate in regulation of very different cellular functions. However, some small GTP-binding proteins can, like the G-proteins, participate in the transduction of external signals to enzymes producing second messengers. The proteins designated as RAS_1 and RAS_2 in yeast participate in transduction of an external nutritional signal from its detector to adenylyl cyclase. When essential nutrients are present adenylyl cyclase is stimulated and the increased concentration of cAMP provides a positive signal for growth. The requirement of RAS proteins for the stimulation of adenylyl cyclase *in vivo* has been demonstrated by genetic experiments. Mutations in RAS which inactivate the function of these proteins cause a block in production of cAMP; the mutant yeast cells have a phenotype similar to that of nutritionally deprived cells. Thus they remain in the resting phase rather than entering a growth and division cycle. The product of mammalian *ras* gene, p21^{ras}, shares considerable sequence homology with yeast RAS. However, the biological role of p21^{ras} is not in the regulation of mammalian adenylyl cyclase. The structure and function of p21^{ras}, which when mutated becomes a potent oncoprotein, will be discussed further below.

2.3 Enzymes that control the concentration of second messengers: structure and regulation

cGMP phosphodiesterase

The interaction between a G-protein and an effector is most clearly defined for that between G$_t$ and cGMP phosphodiesterase in retinal rod cells. The cGMP phosphodiesterase consists of three types of subunit: α (88,000), β (85,000) and γ (9,000). In the absence of light, cGMP phosphodiesterase has very low catalytic activity and the subunits are arranged as an αβγ$_2$ complex. Biochemical studies of the enzyme in this basal state have shown that it can be activated by limited proteolysis. This treatment results in the rapid degradation of only the γ subunit. The remaining αβ complex could be inactivated by the addition of intact γ subunit. These experiments demonstrate that the enzyme consists of a catalytic αβ complex and inhibitory γ subunit and that the activation can be achieved by overcoming inhibitory constraints imposed by γ. Studies of the physiological activation of cGMP phosphodiesterase have shown that the action of G$_t$ employs essentially the same mechanism: α$_t$ subunit, generated by dissociaton of G$_t$, binds to and actively removes the γ subunit from the αβγ$_2$ holoenzyme. The remaining αβ complex is catalytically active and cGMP is rapidly hydrolysed. These data were provided by analysis of protein complexes formed in retinal rod cells after stimulation by light, as well as by detailed kinetic analysis of the interaction between purified G$_t$ and cGMP phosphodiesterase.

Adenylyl cyclase

Activity of mammalian adenylyl cyclase is both stimulated and inhibited through the action of G-proteins. There are several forms of the enzyme. Those purified so far appear to be glycoproteins containing a single polypeptide with a molecular mass of 120-150,000. Determination of the primary structure for one of these enzymes has revealed that this integral membrane protein consists of two large hydrophobic and two hydrophilic domains. It is predicted that each hydrophobic domain contains six membrane spanning α-helices (figure 7). The purified enzyme has been successfully reconstituted with a stimulatory β-adrenergic receptor and G$_s$ into an adrenaline-responsive system. The α$_s$ subunit directly interacts with adenylyl cyclase. The active form of the enzyme is in fact a complex of an adenylyl cyclase molecule and α$_s$ in its GTP-bound form. The components are relatively tightly bound to each other so that the complex can be purified without causing their dissociation.

Inhibition of adenylyl cyclase occurs through stimulation of inhibitory receptors (e.g. α$_2$-adrenergic and the muscarinic M2 subtype) and a G-protein which can be

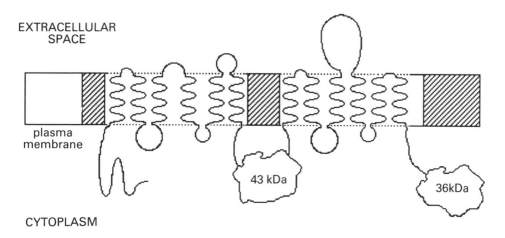

EXTRACELLULAR
SPACE

plasma
membrane

43 kDa

36kDa

CYTOPLASM

Fig 7. Transmembrane topology of adenylyl cyclase. (According to *Krupinski et. al. /1989/ Science 244 : 1558*)

modified by *pertussis toxin*. Reconstitution experiments, however, have not provided a clear demonstration that the inhibitory α subunit (mixture of $\alpha_{i-1,2,3}$) binds tightly to either component of the α_s/adenylyl cyclase complex causing inhibition. Although the molecular mechanism of G-protein inhibitory action remains an open question, one current hypothesis offers the following explanation. Hormone induced dissociation of G_i into α_i and βγ would result in the increase in the concentration of free βγ in the cell. Since βγ could interact with different α subunits, this would then drive interaction with α_s so that it is no longer available for stimulation of adenylyl cyclase. As a result, adenylyl cyclase would be inhibited indirectly.

Phospholipase C

The hydrolysis of PIP_2 to IP_3 and DG can be catalysed by members of a family of enzymes collectively called phosphoinositide-specific phospholipase C (PI-PLC). Several forms of the enzyme have been identified in mammalian cells as well as from invertebrate sources (e.g. *Drosophila melanogaster*). The primary structures of many purified forms have been determined (figure 8). PI-PLC activities consist of a single polypeptide with a molecular mass within a wide range from 60,000 to 150,000. As discussed previously, PI-PLC activity can be stimulated by a number of different agonists. Analysis of such systems has been aimed at the identification of a PI-PLC form participating in a specific signal transduction pathway and the mechanism leading to its activation. For example, one structurally defined form isolated from *Drosophila* (the product of the *norp* A gene) has been identified as a component of a phototransduction system (in some invertebrates, unlike mammalian rod cells, the visual signal is transmitted through generation of IP_3 and DG). In this case, use of a genetic approach has enabled the assignment. Mutations in the gene encoding the PI-PLC that destroyed enzyme activity caused a drastic reduction in the light-evoked responses. Biochemical analysis of phototransduction in wild type flies demonstrated that this form of PI-PLC became activated by the action of a GTP-binding protein.

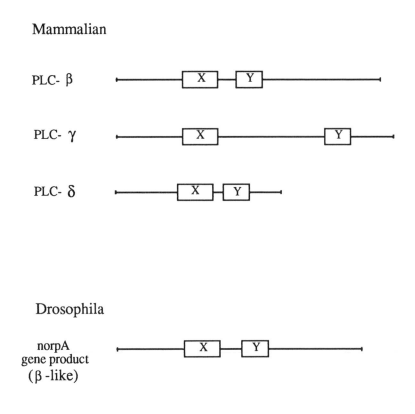

Fig 8. Family of phospholipase C molecules specific for inositol phospholipids. Comparison of amino acid sequences of different PLC forms shows low overall homology. However, two regions present in PLC-β, PLC-γ, and PLC-δ share significant sequence similarity (about 50% of identical amino acid residues). These regions, designated as X and Y, may underlie catalytic function. All forms of PLC presented here, appear to be peripheral membrane proteins. (According to *Rhee et. al. /1989/ Science 244 : 546*) .

G-protein activation of mammalian PI-PLC has been demonstrated in a number of systems. In some cases, the G-protein has been identified as a *pertussis toxin* substrate (e.g. stimulation of neutrophils by the chemotactic peptide fMet-Leu-Phe). However, G-protein mediated stimulation of PI-PLC does not seem to be an exclusive mechanism of regulation. A possibility that some forms of PI-PLC could be distinguished by different regulatory mechanisms will be discussed below.

3. Other transmembrane signalling mechanisms that control the concentration of second messengers

The general mechanism described above is considered to be the predominant mechanism for translating extracellular signals into changes in the concentration of a second messenger. However, in addition to this G-protein mediated regulation of second messenger concentration, other mechanisms have been discovered. At present, they can be illustrated by only a few examples.

One alternative strategy is seen in the stimulation of cGMP production in sea urchin sperm. The extracellular signal is a peptide secreted by sea urchin eggs with the biological role of attracting sperm. Purification of the membrane bound form of guanylyl cyclase demonstrated that the enzyme activity and the receptor function are properties of one polypeptide chain. Analysis of the primary structure of this protein predicted transmembrane, extracellular and intracellular domains. The extracellular domain seems to be related to the receptor function while the intracellular part confers guanylyl cyclase activity. Thus, the activity of guanylyl cyclase can be stimulated through the conformational change of the molecule initiated by the ligand binding (see also Chapter 3).

In several types of mammalian cells, natriuretic peptides (hormones involved in the regulation of fluid and electrolyte homeostasis) stimulate production of cGMP. Analysis of the receptors for natriuretic peptides and guanylyl cyclase in such mammalian cells showed that, as above, these two functions can be contained within one molecule.

Studies of cell proliferation in response to EGF and PDGF suggest another mechanism that controls second messenger concentration. In some mammalian cells stimulation by EGF and PDGF results in rapid hydrolysis of PIP_2. As described in the previous chapter, receptors for these growth factors possess tyrosine kinase activity. One class of inositol-specific phospholipase C (PLC- γ in figure 8) can associate with and be phosphorylated by these receptor kinase activities. Thus the proposed mechanism for the stimulation of phospholipase C activity by EGF and PDGF is phosphorylation of this specific form of PI-PLC by the receptor tyrosine kinase.

4. Second messengers and the regulation of cell growth

Two of the several different mitogenic pathways that have been suggested appear to be mediated by second messengers. As discussed in Chapter 1, no one pathway seems to be obligatory, but effective mitogenesis often requires cooperation between different pathways. Table 1 lists those external signals whose regulation of cell growth involves second messenger pathways (i.e. changes in concentration of cAMP or DG and IP_3 or both) as a part of a complex mitogenic response.

Some external signals that are involved in the regulation of cell growth cause hydrolysis of PIP_2 within minutes after the cells are stimulated; these signals include bombesin, vasopressin, thrombin, angiotensin II as well as PDGF and EGF (in some cell types). Many of the listed external signals are likely to stimulate phospholipase C activity via a GTP-binding protein. PDGF and EGF, however, may employ a different mechanism; the tyrosine kinase of their receptors may phosphorylate and thus stimulate phospholipase C activity.

The role of cAMP in the regulation of cell growth has been demonstrated in several systems. For example in the yeast *S.cerevisiae*, increased concentration of cAMP provides a signal for growth in response to a nutrient supply. In mammalian systems, such a role of cAMP as a positive regulator of growth is well documented only for a small subset of differentiated cells. For example, β-adrenergic agonists, through the increase of cAMP concentration, stimulate proliferation of rat parotid cells. Stimulation of proliferation of thyroid epithelial cells by thyrotropin and pituitary somatotrophs by growth hormone releasing factor also provides examples of this pathway. Stimulation of adenylyl cyclase in those mammalian cells is mediated by G_s. However, in other mammalian cell types (e.g. fibroblastic or tumour cell lines) cAMP has often been regarded as a negative signal for proliferation. In some cases, it has been shown that cAMP elevating agents inhibit growth or that cAMP levels in those cells inversely correlates with proliferation.

5. Proteins encoded by *mas, ras, gsp,* and *crk* oncogenes

Several oncogene products have been directly related to proteins involved in second messenger production. These include the gene products of *mas*, *ras*, *gsp* and *crk*. The *mas* gene product is a member of the seven transmembrane class of receptors and appears to be coupled to PI-PLC activation (see Chapter 3). The other transforming proteins function downstream of receptors and are discussed seperately below.

5.1 An amino acid substitution in the GTP-binding protein p21^{ras} generates a potent transforming protein

The most common oncogenes isolated from human tumours or transformed cells lines are derived from the *ras* gene family. In the human genome there are three distinct *ras* genes (c-Ha *ras*, c-Ki *ras* and c-N *ras*); the products of these genes are the p21^{ras} proteins. Several properties of the p21^{ras} proteins indicate that they may function as coupling proteins between mitogen receptors and effector molecules:

1) The p21^{ras} proteins are located on the cytoplasmic side of the plasma membrane.

2) The ras proteins undergo GTP-induced conformational change (p21^{ras} -GDP inactive, p21^{ras}-GTP active).

3) In yeast cells, p21^{ras} can replace yeast RAS in activation of adenylyl cyclase.

However, the coupling role for p21^{ras} in mammalian cells has not been identified. The only protein known to interact with p21^{ras} is the protein named G̲TP-ase a̲ctivating p̲rotein (GAP). The interaction between these two proteins stimulates the GTP-ase activity of p21^{ras}.

One fact, however, is not disputed: mutations that interfere with p21^{ras} function can lead to malignant transformation. The normal function of this protein is required for the controlled transduction of a mitogenic signal. The difference between a normal *ras* and a transforming *ras* gene is usually a single base change resulting in a single amino acid substitution in the protein. For example, p21^{ras} isolated from many human tumours has the glycine at position 12 replaced by another amino acid residue. Several other amino acid substitutions can also activate the transforming potential of p21^{ras}. In figure 9, the positions of such mutations are indicated within a section of the 3-dimensional model for the structure of the normal protein. Mutations are located within three critical regions related to the functions of guanine nucleotide binding and hydrolysis of GTP. Some mutations interfere with guanine nucleotide binding and facilitate exchange of GDP for GTP. Other mutations can modify the catalytic site of p21^{ras}. As a result of this modification, GTP-ase activity is no longer stimulated through the interaction with GAP. In both cases, mutations lead to accumulation of p21^{ras} in its active, GTP-bound form. This prolonged activation of mutated p21^{ras} can act as a mitogenic signal and ultimately cause cell transformation.

5.2 The putative oncogene *gsp* encodes α_s with a defective GTP-ase activity

A subset of human pituitary tumours carry oncogenic mutations in the gene for the α subunit of G_s, the stimulatory regulator of adenylyl cyclase. These mutations have been identified as amino acid substitutions at two positions important for α_s function. One amino acid substitution occurs within the guanine nucleotide binding region (Gln227) conserved among GTP-binding proteins. Mutation at the corresponding position in p21^{ras} (Gln61) produces a protein that promotes malignant transformation. The other position that can be mutated (Arg201) lies within the region presumed to be important for the high intrinsic rate of GTP hydrolysis (the rate of GTP hydrolysis by α subunits of the heterotrimeric G-proteins described above does not depend on the interaction with other proteins such as GAP as seen in the case of p21^{ras}). The mutations at both positions constitutively activate α_s by inhibiting its GTP-ase activity. The consequent activation of adenylyl cyclase and continuous synthesis of cAMP bypasses the requirement of the pituitary cell for growth hormone releasing factor (GHRF). The GHRF normally provides a positive signal for both cell proliferation and the secretion of growth hormone. Thus, the mutations in α_s promote tumour growth by inducing an autonomous proliferative signal, through the increase in cAMP concentration in pituitary cells.

The final proof that the mutated gene for α_s is an "oncogene" requires a test for the induction of tumours after deliberate expression of the gene. Since the expected tumourigenic effect would be exerted only in a relatively small subset of cells that

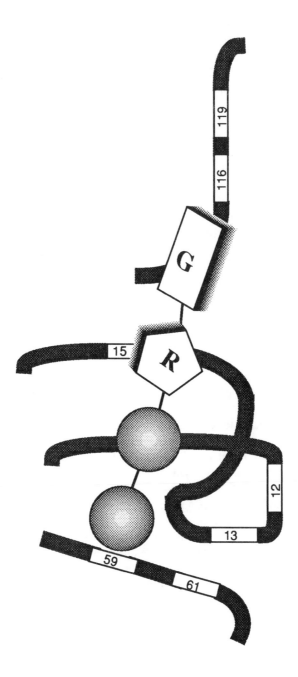

Fig 9. Oncogenic mutations in p21^{ras}. X-ray crystallography was used to resolve the three-dimensional structure of p21^{ras}/GDP complex. The schematic drawing presented here outlines sections of the protein backbone surrounding the bound GDP. The site of mutations that can activate p21^{ras} are indicated by the numbered boxes.

proliferate in response to elevated cAMP, the experiment is not easy to perform. The postulated oncogene is designated *gsp* (gasp), for G̲s̲ protein.

5.3 The protein encoded by *crk* shares structural similarity with several proteins involved in the regulation of cell growth

The *crk* oncogene has been isolated from one of the acutely transforming retro-viruses. The gene product, p47*gag-crk*, is a fusion protein made up of the viral *gag* protein and the cellular protein encoded by the proto-oncogene *crk* (figure 10). Within the amino acid sequence of p47*gag-crk* there are two regions (SH2 and SH3) that are homologous with sequences of several cellular proteins. These cellular proteins (e.g. phospholipase C, non-receptor tyrosine kinases and GAP) are implicated in the regulation of cell growth. Both, SH2 and SH3 regions are present in one form of phospholipase C and most families of non-receptor tyrosine kinases. In the latter, SH2 and SH3 regions are transposed relative to their order in p47*gag-crk* and phospholipase C (see figure 10). Some cellular proteins contain only one of these two regions. A protein of the cytoskeleton, α-spectrin, contains a single SH3 region and the protein that activates the GTP-ase activity of p21*ras*, GAP (see above), has two SH2 regions.

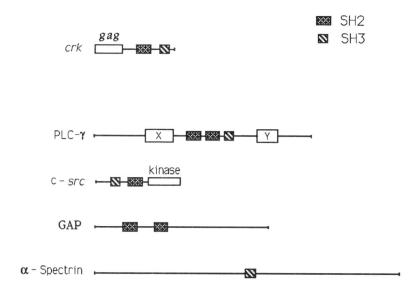

Fig 10. Homology between p47*gag-crk* and cellular proteins. The primary structure of p47*gag-crk* shares sequence similarity with PLC-γ, non-receptor tyrosine kinases (e.g. the kinase encoded by *c-src*), GAP (GTP-ase activating protein of p21*ras*) and α-spectrin. The regions of homology are designated as SH2 (chequered) and SH3 (hatched).

The regions SH2 and SH3 are not related to any known catalytic activity. They might, however, represent regulatory regions that permit interaction with other cellular proteins. Binding of such proteins to SH2 and SH3 regions of phospholipase C or tyrosine kinases could influence the catalytic activity contained within the separate domains. The SH2 and SH3 regions in p47$^{gag\text{-}crk}$ could bind the same regulatory proteins and make them unavailable for their normal functon. According to the current hypothesis, p47$^{gag\text{-}crk}$ could bind proteins that inhibit function of phospholipase C, tyrosine kinase, or GAP and thus cause their activation. This break in the control of such important regulatory proteins could lead to the malignant transformation induced by p47$^{gag\text{-}crk}$.

5

Protein kinases

1. Protein phosphorylation

The previous chapter discussed the means by which receptors operate through the production of second messengers. Logically then, the next issue to be addressed in signal transduction is the nature of the targets for these second messengers. The response of a cell to a simple nutrient such as glucose is effected when the sugar itself is transported into the cell and signals its presence through direct binding to proteins, so inducing a shift in energy usage. In contrast responses to hormones, growth factors or neurotransmitters, in multicellular organisms are necessarily more complex. It is perhaps not surprising then that what is known of the targets of hormone induced second messengers places many within a superfamily of regulatory proteins (the protein kinases) that each act on multiple cellular events and so coordinate an appropriate cellular response.

Protein phosphorylation which is catalysed by these regulatory enzymes is one of a number of post-translational modifications that are employed by cells; others include acetylation, ribosylation, glycosylation, acylation and ubiquitinylation. Each of these modifications serves a role in controlling protein function or in directing protein localization within the cell. In the case of protein phosphorylation however, there is abundant evidence for a role in the dynamic control of cellular events. Thus phosphate esters bonds on target proteins are repeatedly made and broken with consequent activation/inactivation cycles of the target protein. This means of control is ubiquitous and operates across the evolutionary divide from prokaryotes to man.

The protein kinases catalyse the transfer of the γ-phosphate of ATP (or GTP) to an hydroxyl group on an amino acid side chain of serine, threonine or tyrosine residues within a polypeptide. The addition of a phosphate group to these amino acids results in the introduction of a negative charge at that site and is generally thought to alter conformation of, or accessibility to, the phosphorylated protein. Perhaps the clearest example of such changes is afforded by the most extensively studied phospho-protein, phosphorylase a. This protein was in fact the first to be shown to be regulated through phosphorylation, the phosphorylated form (phosphorylase a) being active. Phosphorylase b is converted to the a form through the action of phosphorylase kinase which phosphorylates a serine residue close to the amino terminus. From crystal structures of phosphorylase, it is evident that this amino-terminal stretch is displaced following phosphorylation, so permitting phosphorylase to function. The regulation of this enzyme by phosphorylation illustrates a number of important points with respect to phosphorylation control. The site of phosphorylation is and indeed must be accessible to the enzymes responsible for the transfer and hydrolysis of the phosphate ester group. The removal of phosphate permits the enzyme to return to an inactive state, so the control operates in a dynamic fashion and obviates the need for protein synthesis in controlling cellular processes. Lastly, control of phosphorylase by phosphorylation occurs in addition to allosteric control; the latter allows the cell to maintain what is in this case a basic function without the influence of exogenous factors.

The control of phosphorylase by phosphorylation is in some respects a simplified system in that a single phosphorylation site is involved. In contrast many phosphoproteins contain multiple sites of phosphorylation and are controlled by the action of multiple protein kinases (and possibly more than one phosphoprotein phosphatase). This multisite phosphorylation permits the integration of responses to different agonists which physiologically are not presented to cells one at a time. The complexities of the integration and networking of signal transduction pathways are covered in a separate chapter (Chapter 8) as are the nature and regulation of the phosphoprotein phosphatases (Chapter 6).

This chapter will describe the protein kinase superfamily of regulatory proteins. In particular emphasis will be placed on the protein kinases that are regulated through second messengers and therefore play a defined role in a signal transduction cascade.

2. The diversity of the protein kinase superfamily

There are at present in excess of 110 unique gene products that make up the protein kinase superfamily. Functionally these have been subdivided on the basis of specificity with respect to the target amino acid, i.e. there are protein tyrosine kinases and protein serine/threonine kinases (there are also activities which will phosphorylate other amino acid residues but little is known about these activities or consequences of these phosphorylations). No individual polypeptide has been found to be able to catalyse the phosphorylation of both tyrosine and serine/threonine residues (this contrasts with certain of the protein phosphatases that can dephosphorylate tyrosine as well as serine/threonine residues; see Chapter 6). This functional distinction parallels structural distinctions which have been elucidated from the determination and comparison of primary amino acid sequences. Alignments of the catalytic domain sequences of all known protein kinases allows them to be placed within an evolutionary tree based upon sequence divergence (figure 1). This analysis of similarity shows that the protein tyrosine kinases form a distinct branch of this family. Furthermore, within the tyrosine kinase branch there are subdivisions which relate to the receptor class of protein tyrosine kinases (see Chapter 3) and the so called non-receptor class (see below). Similar subdivisions which also relate to function are evident within the protein serine/threonine specific kinases (figure 1). These include for example the cyclic nucleotide-dependent protein kinases that are part of particular signal transducing pathways as discussed further below.

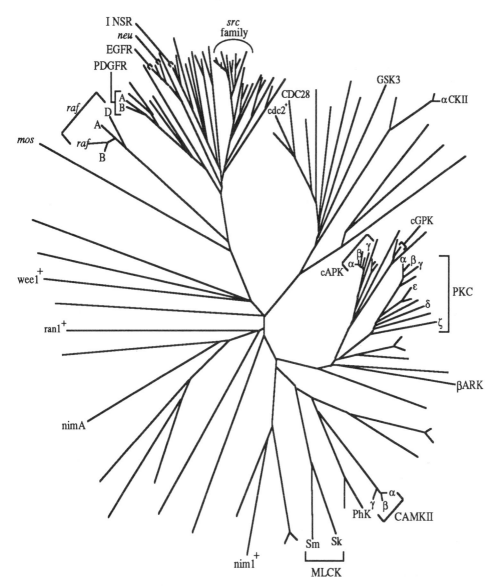

Fig 1. The protein kinases are linked according to the protein sequence conservation of their catalytic domains. The end of each branch denotes a specific protein kinase. Those named include (i) certain members of the receptor tyrosine kinase class PDGFR, EGFR, neu and InsR discussed in Chapter 3 (ii) kinases involved in cell cycle control in yeast and Aspergillus, wee1[+], ran1[+], nimA, nim1[+], cdc2[+] and CDC28 (iii) Ca-CAM-dependent enzymes, MLCK, PhK and CAMKII (iv) βARK responsible for homologous desensitization (see Chapter 8) (v) members of two kinase families cAPK and PKC. Branches closely associated with for example the PKC (rat) family are related sequences from other species (Drosophila, nematode). The closely related cluster of proteins including the PDGFR and the *src* family of kinases represent the protein tyrosine kinase class. (Courtesy of Dr S. Hanks, Vanderbilt University School of Medicine, Nashville, U.S.A.)

It is possible to generate such a family tree of protein kinases because there is within the primary sequences of all of these polypeptides a domain that is conserved (it is the alignment of this domain that provides the family tree). This domain, which is in essence the catalytic core of these proteins, can occur essentially anywhere within the linear sequence; for example in the protein encoded by *c-abl* it is at the amino-terminus, for protein kinase C it is at the carboxy-terminus, while for the cAMP-dependent protein kinase it spans virtually the entire polypeptide chain of one subunit (the catalytic subunit; see below).

Although these conserved catalytic domains are easily recognizable within a polypeptide sequence, there are only a few invariant residues that are present in all of these protein kinases (see figure 2). These residues are presumed to play a fundamental

```
..GXGXXG....AhK.........................DL...DFG........APE......DXWSXG.................
```

Fig 2. Single letter code amino acid residues are shown, indicating those that are highly conserved between all members of the protein kinase superfamily. X denotes any residue and h denotes a hydrophobic residue.

role in the binding of Mg-ATP and in the transfer of the γ-phosphate residue from the nucleotide triphosphate to the acceptor hydroxyl group on the target amino acid side chain. An indication that this is indeed the case comes from crystallographic studies on other nucleotide binding proteins. For example the *ras* encoded proteins discussed in the previous chapter contain a glycine rich stretch which forms a pocket enclosing the three phosphate groups of the bound Mg-GTP. A similar GXGXXG sequence is found in all of the protein kinases[1]. Within 20 amino acids of this glycine rich sequence is an invariant lysine residue which can be cross-linked with a photo-activatable ATP analogue again suggesting proximity to the ATP binding site. That this lysine is necessary for the function of the kinases is evidenced by mutagenesis of this residue. Replacement of this lysine with any other residue leads to the expression of an inactive kinase (figure 3). This not only provides an indication of involvement in catalytic function, but underscores the extreme evolutionary pressure on the maintenance of this invariant lysine.

It is not surprising given the pattern of the family tree in figure 1 that there are distinct conserved characteristics that separate the protein serine/threonine kinases from the protein tyrosine kinases; for example the DLKPEN sequence shown in figure 4.

[1] One of the glycines is replaced in certain protein kinases.

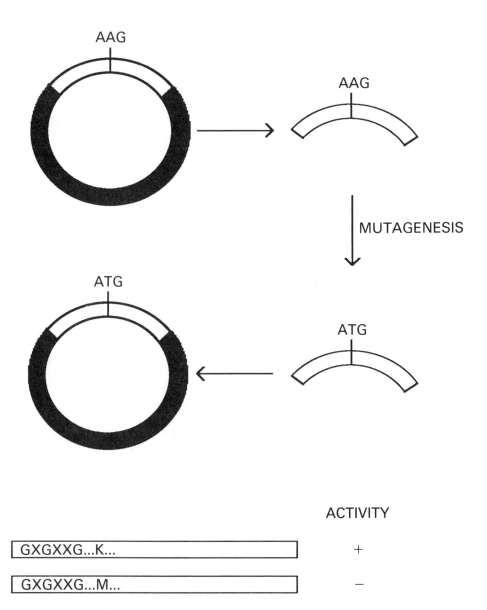

Fig 3. The cDNA for a protein kinase (open sector) is carried in an appropriate plasmid (filled sector). Excision of the cDNA, usually a fragment, and use of site directed mutagenesis permits the selective alteration of the coding sequence; in this case an A to T change that alters the coding triplet for lysine (K) to methionine (M). This altered sequence can then be replaced in the original vector. Expression of the encoded proteins containing wild type or mutant kinase demonstrate that while both can be expressed the alteration to this conserved lysine abolishes activity. (It should be noted that some introduced mutations may lead to expression of, for example, unstable but nevertheless active proteins.)

These distinctions permit the classification of protein kinases into serine/threonine or tyrosine kinases without prior knowledge of function. This has been particularly useful since a number of putative protein kinases have been defined only through identification of the genes that encode them, as opposed to isolation of a functional protein. Initially kinases were isolated as proteins based upon their ability to phosphorylate specific substrates. However more recently kinases have been identified using oligonucleotide or cDNA probes i.e. on the basis of homology to previously identified polypeptides. Present understanding of this protein family means that on the basis of the particular pattern of conserved sequences it is possible to classify such proteins as either serine/threonine or tyrosine kinases.

serine/threonine specific

```
..GXGXXG....AhK.....................DLKPEN..DFG......APE......DXWSXG...............
```

tyrosine specific

```
..GXGXXG....AhK.....................HRDL..K.DFG...P.W.APE...DXWSXG...............
```

Fig 4. Conserved residues in the catalytic domains of these two subfamilies of the protein kinases are shown. Single letter amino acid code is used; X denotes any residue, h denotes hydrophobic residue.

3. Serine/threonine-specific protein kinases

There are many more protein kinases than there are defined signalling pathways. As such there are a number of what can be termed orphan kinases for which the input signal is essentially unknown or at least not traceable to the cell surface. For the purposes of discussion the second messenger-dependent protein kinases, which fit clearly into particular signal transduction pathways, and these orphan kinases will be considered separately.

3.1 Second messenger-dependent protein kinases

The second messenger-dependent protein kinases can be subdivided on the basis of the nature of the second messenger itself, i.e. cyclic nucleotide, diacylglycerol or Ca^{2+}-dependent. The kinases associated with these second messengers are discussed separately below.

Cyclic nucleotide-dependent protein kinases

The cyclic AMP-dependent protein kinase (cAPK) was the first broad specificity kinase to be identified. Two forms of this enzyme have been separated on the basis of elution from ion exchange columns (see figure 5a). Purification of these proteins has led to the identification of two classes of regulatory subunits, RI associated with

(a)

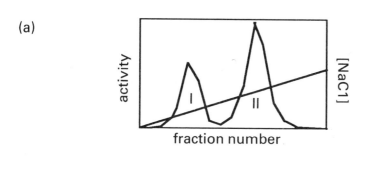

(b)

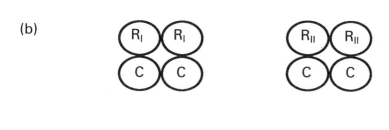

(c)

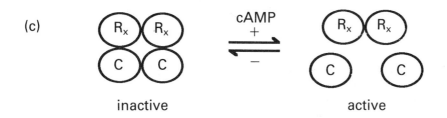

(d)

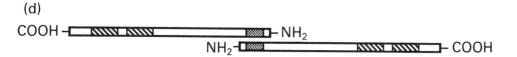

Fig 5. (a) Chromatography of extracts on an anion exchange column seperates two kinase activities that are dependent upon cAMP for activity. These are termed peaks I and II.

(b) Purification of peaks I and II leads to the isolation of two heterotetramers. The peak I yields a regulatory (R) subunit denoted R_I while the second peak yields R_{II}; the catalytic subunit (C) is common to both.

(c) Activation of the cAPK proceeds through the binding of cAMP to the R subunits. This leads to dissociation of the holoenzyme and the released monomeric C subunits that are active.

(d) The domain structure of the R subunit dimer includes two cAMP binding sites (hatched boxes) and a dimerization domain required for R-R interactions (stippled boxes).

peak I and RII associated with peak II; the catalytic subunit (C-subunit) is common to both (5b). As indicated in figure 5c the holoenzyme is inactive; following the binding of cAMP to the R subunits (2 sites/R subunit) the C-subunit dissociates as an active monomer. The expression of this kinase activity is responsible for virtually all the effects of elevated cAMP.[2]

Investigation of phosphorylation sites in substrates of the cAPK has revealed a particular pattern of basic residues that has been further defined through the use of synthetic peptide substrates. Thus for the cAPK there is a linear sequence requirement for a substrate consisting of a dibasic sequence amino-terminal to the target serine/threonine i.e. R/K R/K X S/T. This primary structural requirement is of use in predicting potential cAPK phosphorylation sites, particularly in conjunction with secondary structure prediction. In order to be accessible to protein kinases all sites must be on the surface of a protein and indeed they are frequently found at presumptive β-turns.

A second aspect of the primary structural specificity is that it has been used in generating inhibitors (with the target serine/threonine residue replaced by alanine which has no side chain hydroxyl group). Synthetic peptides based on the sequence R/K R/K X A act as inhibitors of the cAPK. This understanding of peptide recognition sequences has also contributed to defining how regulatory subunits (or domains) interact with and inhibit the catalytic subunit (or domain). All the second messenger-dependent protein kinases require by definition the presence of a second messenger for activity. Thus as synthesized they are locked in an inactive state and the question then arises as to how the activity is suppressed. In the case of the cAMP-dependent protein kinase it is now generally accepted that at the site of RI interaction with C there is a so called pseudosubstrate site that has the hallmarks of a phosphorylation site (i.e. a substrate site) but no target residue (serine/threonine); peptides based on this RI sequence are inhibitory. It appears then that inhibition of C is effected by interaction of the substrate recognition site with this pseudosubstrate site in RI; mutation of a similar site in the RII subunit is consistent with this model for R-C interactions and activity suppression. The conformational change(s) associated with cAMP binding to the R(I or II) subunits is responsible for reducing the avidity of the R-C interaction, leading to dissociation and release of the catalytically active C subunits. Similar substrate or pseudosubstrate sites occur in other protein kinases and the example described here serves as a general model for second messenger-dependent kinase regulation.

In addition to the pseudosubstrate inhibitory domain discussed above the R subunits also have cAMP binding sites. An understanding of the structure of these sites has been significantly aided by the crystal structure of the bacterial protein CAP, which like the RI and RII subunits binds cAMP (in the case of the CAP protein however it functions as a cAMP regulated DNA binding protein). Thus in both RI and RII there are two sequences related to CAP; these form a tandem repeat defining two homologous but non-identical cAMP binding sites (figure 5d); these sites can in fact be distinguished through the use of cAMP analogues, i.e. they are not functionally identical. Between the cAMP binding domain and the amino-terminus lies the site

---

[2] A cAMP gated ion channel has also been described.

of interaction with the C-subunit (pseudosubstrate site) and beyond this the domain responsible for R-R dimerization (figure 5d).

The further definition of the structure/function of C and RI/II subunits will no doubt come from mutagenesis and expression of cDNAs for these proteins. To date the isolation of cDNAs has already provided evidence of diversity with at least two genes for each of the C, RI and RII subunits. The functional consequences of the expression of different combinations of these gene products have yet to be elucidated.

The involvement of the cAPK in regulatory events can be inferred from the use of cAMP analogues that are readily membrane permeable; both dibutyryl cAMP and 8-bromo cAMP are frequently used in this context.[3] However to define a physiological target for cAPK (or any other kinase) it is generally accepted that certain criteria are met; these can be generalized as follows:

1. The target protein is phosphorylated *in vitro* by the cAPK (or other PK).

2. Activators of the cAPK (or other PK) stimulate the phosphorylation of the target protein *in vivo*.

3. The site(s) of phosphorylation *in vitro* and *in vivo* is the same.

4. Where applicable, the effect of phosphorylation of the target protein should reflect the physiological response (e.g. elevated cAMP reduces the rate of glycogen synthesis in muscle; when cAPK phosphorylates glycogen synthetase it reduces its activity).

In many instances at least the first three criteria have been met and the list of physiological cAPK substrates is substantial, many being involved in the regulation of cellular metabolism. However with respect to the cAPK and cell proliferation no specific targets of consequence have been defined. Nevertheless the ability of the cAMP pathway to promote mitogenesis is clear and exogenous cAMP analogues or cAMP elevating agents (e.g. *cholera toxin*, forskolin, see Chapter 4) will act alone to cause pituitary cell proliferation (see *gsp* Chapter 4) and in synergy with other factors in the proliferation of many other cell types. Responses are not always predictable and proliferative responses are not universal; in a number of cell types elevation of cAMP induces differentiation and cessation of division. The complex pattern of cell specific responses to cAPK activation is a feature of other signal transduction pathways as discussed further in Chapter 8.

In many respects the cGMP-dependent protein kinase (cGPK) resembles the cAPK. Structurally the cGPK appears to consist of two R-C "fusion" proteins (see figure 6) and it has in fact been proposed that the cGPK gene evolved by gene fusion from ancestral genes common to the cAPK. Structural work on the cGPK has led to the

---

[3] It is worthwhile noting that the use of analogues is not always ideal. Thus dibutyryl cAMP is metabolized to generate AMP and butyrate. Butyrate in its own right is a potent enhancer of differentiation in a number of systems. Thus dibutyryl cAMP can produce effects that are not attributable to the activation of cAPK.

identification of two splice variants of the cGPK. The alternative splicing involves the first ~100 amino acids, covering the dimerization region of the protein. The significance of this biological variation has yet to be defined.

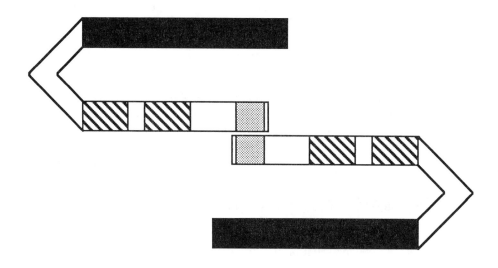

Fig 6. Domain structure of the cGPK. A schematic domain structure of the cGPK dimer is shown. Indicated are the dimerisation domain (stippled), cGMP binding domain (hatched) and catalytic domains (black).

The cGPK has a substrate specificity that is similar to cAPK; however selective substrates have been identified. For example the G-substrate as its name implies is an abundant cGPK substrate expressed in Purkinje cells in the brain. As with cAPK the role of the cGPK has been addressed through the use of membrane permeant cGMP analogues and agents that artificially cause elevation of endogenous cGMP concentrations. An example of cGPK involvement in hormone action is provided by the atrial natriuretic peptide receptors which encode ligand activated guanylyl cyclases as discussed in Chapter 3. Cellular responses to this class of hormones appear to be mediated through the activation of cGPK.

Diacylglycerol regulated protein kinase C

Protein kinase C (PKC) was originally described as a protein kinase that was activated by the neutral lipid diacylglycerol in the presence of Ca^{2+} and anionic phospholipid (figure 7a). Diacylglycerol (abbreviated as DG or DAG) is produced by PLC hydrolysis of phospholipids in response to a number of agonists as described in the previous chapter; as such it is proposed that PKC plays a primary role in signal transduction from these agonists (figure 7b).

[a]

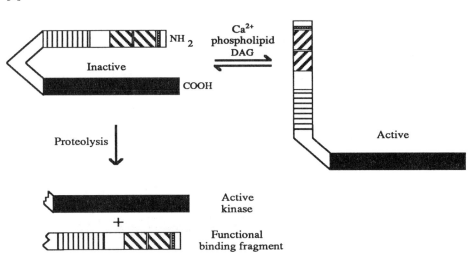

[b]

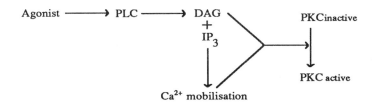

Fig 7. Shown in [a] is the domain structure of protein kinase C (PKC) with the two DAG/phospholipid binding sites indicated by the hatched boxes; the kinase domain is the black boxed area and the pseudo-substrate site (see text) is the stippled box near the amino-terminus. The striped box represents a domain of unknown function that is conserved in PKC α, β, γ but not δ, ϵ, ζ (see figure 8).

[b] The activation of PKC in response to agonists is effected through the hydrolysis of inositol lipids, ultimately leading to the production of DAG and increased cytosolic Ca^{2+} as indicated.

The primary structure of PKC has been determined through a combination of partial protein sequence and the use of oligonucleotide probes in the isolation of complete cDNAs. Subsequently some related proteins were identified by using low stringency screening[4] of cDNA libraries. These approaches have revealed that what was originally considered to be a single protein/activity is in fact made up of a family of related proteins, the present members of which are shown schematically in figure 8. In all these PKCs there is a catalytic C-terminal domain that can be defined both through homology to all other kinases (as discussed above) and through partial proteolysis (figure 7a). The regulatory domain (N-terminal portion) also retains function following limited proteolysis; it binds diacylglycerol in a phospholipid dependent manner. Unlike the intact protein, the kinase domain generated by limited proteolysis is independent of effectors and is constitutively active; thus in the intact protein the regulatory domain inhibits the kinase activity. There is in all these PKCs a pseudo-substrate site that lies immediately adjacent to the C1 domain (figure 8) and

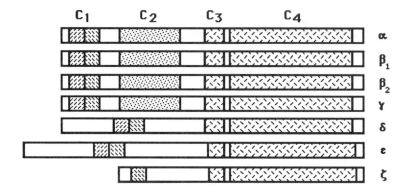

Fig 8. The conserved domains (C_{1-4}) present in most of the currently known PKC gene products are shown. All these PKCs show essentially colinear kinase domains spanning $C_3 + C_4$. By contrast the regulatory domains are more varied with no C_2 domain present in δ, ε or ζ. All these polypeptides do nevertheless contain at least one of the cysteine rich regions present in the first conserved domain (C_1).

this is presumed to be responsible for the inhibition of kinase activity (mutagenesis of a PKC-α cDNA has provided good evidence that this is indeed the case). Interestingly the $-\delta$, $-\varepsilon$, $-\zeta$ PKCs do not have a C2 domain and it transpires that PKC-ε (and probably $-\delta$ and $-\zeta$) is Ca^{2+}-independent. Thus it is surmised that the C2 domain may confer Ca^{2+} dependence.

[4]In screening cDNA libraries, probes are used to hybridize to the inserted cDNAs in order to identify homologous inserts. The stability of the duplexes formed (i.e. probe-cDNA duplexes) is dependent upon the ionic strength and temperature. At low ionic and high temperature (high stringency) only well matched probe-cDNA duplexes will remain; at high ionic strength and low temperature mismatches can be tolerated. In lowering the stringency of the conditions of hybridization and subsequent washing, probes can be used to identify related but non-identical cDNA species.

These PKC proteins show different patterns of expression in cells and tissues; as such it is anticipated that they play similar but distinct roles in signal transduction processes. While there is no clear information yet available to define these roles, we can start to understand their potential through a consideration of the properties and structures of these proteins. There is evidence for a difference in substrate specificity for different PKC isotypes (although there is also considerable overlap in specificity); this suggests that the selective expression of individual PKC isotypes will lead to the targeting of a particular group of proteins. It is also well established that PKC–α, –β1/2 and –γ are Ca^{2+} dependent unlike PKC–ϵ which is Ca^{2+}-independent. The diacylglycerol-dependent activation of PKC–α, –β1/2 and –γ probably requires a coincident rise in cytosolic Ca^{2+} *in vivo*; this is achieved through the hydrolysis of phosphatidylinositol (4,5)bisphosphate(PIP_2) to DAG + IP_3 providing elevated Ca^{2+} through IP_3 triggered release (figure 9). In contrast, although PKC–ϵ would be activated following DAG production from PIP_2 hydrolysis, it would also become active if DAG was produced in the hydrolysis of other lipids (figure 9); there is increasing evidence that DAG is generated from other lipids in particular phosphatidylcholine (figure 9). It is therefore probable that in some tissues the Ca^{2+}-dependent (–α,–β,–γ) and independent (–α–ϵ,–ζ) subclasses of PKC lie on distinct signal transducing pathways.

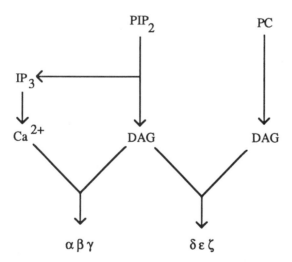

Fig 9. The typical pathway of inositol lipid hydrolysis, leading to both Ca^{2+} and diacylglycerol (DAG) production would be capable of leading to activation of all isotypes. In contrast those that operate independent of Ca^{2+} (δ,ϵ,ζ) could be activated through DAG production that is not coupled to an increase in cytosolic Ca^{2+}.

An important issue with respect to Ca^{2+}-dependent signal transduction pathways concerns how they operate in tissues such as muscle, where the tissue function is intimately involved in the voluntary or involuntary mechanical response to Ca^{2+} movements. How are muscle functions regulated if they are not directly linked to the mechanics of muscle contraction? A simple solution would be to replace those Ca^{2+}-dependent processes with independent ones. This solution appears to have been applied to PKC, since it turns out that PKC–ε (a Ca^{2+}-independent PKC) is expressed at high levels (but not exclusively) in various adult rodent muscle tissues.

As with the cyclic-nucleotide dependent protein kinases, the involvement of PKC in particular responses/phosphorylation events is inferred from the use of membrane permeant activators. These are generally of two types. Firstly, short chain diacylglycerols such as dioctanoylglycerol (two eight carbon acyl side chains) are sufficiently permeant to allow intercalation into cellular membranes. In physiological experiments repeated doses of diacylglycerol are required in monitoring long-term changes because it is rapidly metabolized by the target cell. Secondly and of significance is a group of organic compounds that fall into the phorbol ester class of tumour promoters (tumour promoters do not cause tumours in their own right, but "promote" the subthreshold action of carcinogens, thus causing tumours).

The phorbol ester class of promoters have been shown to activate PKC by mimicking the effect of diacylglycerol. It is supposed that the diverse effects of these agents operate through PKC; similarly the involvement of PKC in a response is often inferred from the ability of a phorbol ester such as tetradecanoyl phorbol acetate (TPA) to produce a response. Both diacylglycerols and phorbol esters have been shown in various cell types to induce many responses and to alter the pattern of cellular protein phosphorylation. In most instances the phosphoproteins induced have not been identified and there is a significant gap in our knowledge of cause and effect. However some *in vivo* substrates for PKC have been defined and several of these are membrane proteins; this is perhaps not surprising given that physiological activation of PKC is dependent upon interaction with diacylglycerol in association with a phospholipid bilayer. It would appear that the phosphorylation of these membrane proteins (such as the EGF receptor, Chapter 3) is associated with desensitization and/or internalization of the proteins. This phenomenon which is discussed further in Chapter 8 can be categorized either as feedback inhibition (in this case where a receptor-PLC pathway is involved) or transmodulation (where a heterologous pathway is involved).

With respect to a role in cell proliferation, there is an abundance of literature indicating that the PKC pathway(s) can contribute. This is evidenced for example in contact inhibited, quiescent Swiss 3T3 cells where in the presence of insulin (which has no affect alone) either diacyglycerol or TPA can induce DNA synthesis. Further evidence for an involvement of PKC in proliferation comes from the observation that many mitogens stimulate lipid hydrolysis with the generation of diacylglycerol, the PKC second messenger. Perhaps the most direct evidence comes from the isolation of a uv-induced mutant PKC that appears to be responsible for transformation of a cell line in culture (discussed below).

Ca²⁺-calmodulin dependent protein kinases

There are various protein kinases that fall into this class. Unlike the cyclic nucleotide-dependent protein kinases or the PKCs this group is structurally heterogeneous and a number show a very limited substrate specificity; the broad specificity of responses to elevated Ca^{2+} may in part be effected through the multifunctional nature of calmodulin itself which has the ability to bind Ca^{2+} and interact with these various kinases (see figure 10). All kinases of this group bind Ca^{2+}-calmodulin (Ca^{2+}-CAM) with high affinity and in the case of phosphorylase kinase calmodulin constitutes the fourth subunit of this multimeric complex. Given the variety of kinases in this class they are discussed briefly below on an individual basis.

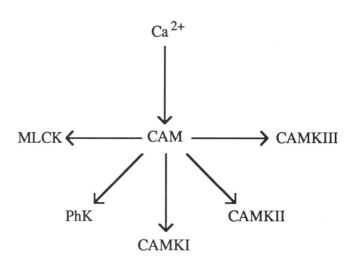

Fig 10. Alterations in Ca^{2+} concentration are monitored in part through the Ca^{2+} binding protein calmodulin (CAM). The Ca^{2+}-CAM complex is itself responsible for the activation of many events, including the protein kinases shown here: myosin light chain kinase, MLCK; phosphorylase kinase, PhK; Ca^{2+}-calmodulin kinases I-III, CAMKI -III.

Myosin light chain kinase (MLCK) was the first Ca^{2+}-CAM dependent kinase identified. As the name indicates MLCK readily phosphorylates myosin light chain and shows a specificity restricted to this substrate. There are two forms of MLCK derived from smooth muscle and skeletal muscle, performing distinct functions in these tissues. In the case of MLCK in skeletal muscle there is evidence that MLC phosphorylation is responsible for modulation of tension during contraction. By contrast in smooth muscle MLCK action appears to be responsible for the initiation of contraction.

Phosphorylase kinase (PhK) is a multimeric complex of α (fast twitch), or α' (slow twitch), β, γ and δ subunits. The δ subunit is CAM itself while the γ subunit is the catalytic subunit. Ca^{2+} mobilization leads to activation through binding to the δ subunit thus permitting phosphorylation of phosphorylase b with a consequent increase in glycogenolysis. PhK can also be activated through phosphorylation by the cAPK (α and β subunit phosphorylation), providing a means of activating glycogenolysis that is independent of Ca^{2+} mobilization. Thus PhK can lead to glycogenolysis through two distinct input signals (Ca^{2+} and cAMP).

There are differences in the properties of skeletal muscle and hepatic PhK which reflects the distinct roles of these enzymes in these tissues. The basis for this functional difference appears to be the expression of distinct genes in these tissues. This is evidenced by the existence of a murine muscular dystrophy characterized by a defective muscle PhK; no apparent alteration occurs in the hepatic PhK. The phenotype of this disease highlights the specificity of PhK genes in their regulatory roles.

Ca^{2+}-CAM kinase I was first identified in brain tissues where it was found to phosphorylate a particular site on the synaptosomal protein synapsin I; it also phosphorylates the neuronal polypeptide protein III. This enzyme is present at high concentrations in the brain.

Ca^{2+}-CAM kinase II is a ubiquitous enzyme that has also been termed the multifunctional Ca^{2+}-CAM kinase because of its broad specificity. This protein is an oligomeric polypeptide variably made up of α (60,000) and β (55,000) subunits in a dodecahedral structure. cDNAs for both α and β have been isolated and expression of either or both together leads to expression of Ca^{2+}-CAM kinase activity; the only difference being that β subunits alone will not oligomerize; a γ gene has also been identified. Substrates for this enzyme include synapsin I, tryptophan hydroxylase, skeletal muscle glycogen synthetase, and the microtubule associated proteins Tau and map-2. As discussed above for other kinases, Ca^{2+}-CAM kinase II has a pseudosubstrate site (in each subunit) that is responsible for the inactivity of the holoenzyme. This inhibitory site is located C-terminal to the catalytic domain and immediately N-terminal to the CAM binding site. The juxtaposition of the CAM binding site and the inhibitory site is essential for the relief of inhibition afforded by CAM binding.

One unusual feature of Ca^{2+}-CAM kinase II, is that following activation by Ca^{2+}-CAM, the enzyme converts to a Ca^{2+}-CAM independent form as a consequence of autophosphorylation. Thus in this case the reversal of the effect of Ca^{2+} release is not simply controlled by the rate of Ca^{2+}-sequestration and dephosphorylation of the target protein(s), but also by the half-life of the phosphates in the autophosphorylated kinase itself. It has been proposed that such mechanisms operate in neuronal memory processes.

Ca^{2+}-CAM kinase III appears to be responsible for regulation of the translational elongation factor EF-2. Phosphorylation of this protein *in vivo* or *in vitro* leads to inhibition of protein synthesis. The role that this kinase plays in negative or positive growth control processes remains to be established.

In contrast to the various PKCs discussed above, it would appear that Nature has in part chosen to diversify her repertoire of responses to Ca^{2+} through the single, highly conserved protein calmodulin (CAM) acting on multiple distantly related regulatory protein kinases. Thus although these kinases appear structurally unrelated they are tied together as a group through interaction with Ca^{2+}-CAM and function to integrate responses to Ca^{2+} mobilization.[5]

3.2. Other serine/threonine-specific protein kinases

There are many serine/threonine-specific protein kinases other than those covered in section 3.1. Aspects of the regulation of some of these is understood. For example, the ubiquitous enzyme casein kinase II (CKII; so called because of its *in vitro* substrate specificity) is a heterotetramer consisting of two α and two β subunits. CKII can be activated by polyamines (spermine or spermidine) and inhibited by heparin. However the physiological relevance of these effects and the role that this enzyme plays in signal transduction is not clear[6]. This partial understanding is typical of the current state of affairs with respect to a number of other kinases.

In the case of the kinase GSK-3 (glycogen synthase kinase-3) there are a number of proteins that appear to be substrates for this kinase *in vivo* and for which there is some physiological 'sense'. The apparent roles for GSK-3 include two opposing events, namely the phosphorylation and inactivation of glycogen synthase and the activation of the type1 phosphatase-modulator complex (further discussed in Chapter 6). Precisely how the effects of this kinase are balanced and indeed how it is regulated with respect to hormone action remains to be elucidated.

While it is likely that further second messenger systems will be uncovered, it is not the case that all the orphan kinases will be regulated directly through the action of second messengers. Rather a number (perhaps the vast majority) of these kinases will form elements in a cascade of kinases much as the prototype PhK. It is already clear that such cascades operate in situations other than glycogenolysis. For example there is a cascade of kinases that appear to be responsible for the coordinate control of acetylCoA carboxylase and HMGCoA reductase. Some of these cascades no doubt originate in second messenger systems or in receptor tyrosine kinases.

---

[5] In fact one can generalize to include other Ca^{2+} binding proteins. Thus responses to Ca^{2+} are in effect perceived through not only CAM but a number of related specialized Ca^{2+} binding proteins such as troponin C and also through distinct families of Ca^{2+} binding proteins such as the 'lipocortins'. In this light, one can conclude that the response of cells to Ca^{2+} is transduced through multiple families of regulatory Ca^{2+} binding proteins.

[6] There is evidence that CKII becomes activated in response to insulin. Whether this reflects a kinase cascade from the insulin receptor tyrosine kinase to this serine/threonine kinase remains to be established.

4. Receptor and non-receptor protein tyrosine kinases

Of the tyrosine kinase class of protein kinases, the receptor class was discussed in detail in Chapter 3 and will be discussed only briefly here. The non-receptor class which represents the much larger class of these kinases is considered in more detail.

4.1 Receptor protein tyrosine kinases

The receptor tyrosine kinases clearly have a very definite 'home' in one or other signal transduction pathway as defined by a ligand. This group of cell surface proteins can be considered to bypass the need for complex receptor coupling mechanisms in that they operate directly as ligand activated kinases. Thus with respect to the generic transducer, the input signal is clearly defined. In this context, there is a significant accumulation of information concerning modulation of the function of these receptors and this is covered in Chapter 8. What is lacking in our understanding of the role of these proteins is the effect of the output. It has been known for more than a decade that these receptors have the potential to phosphorylate proteins, but to date there is no clear instance of such a phosphorylation having a defined physiological effect. Perhaps the reason for this gap in our understanding is simply that these receptor tyrosine kinases are not directly responsible for altering the function of a cellular enzyme that is itself responsible for a measurable event. Nevertheless some potential substrates have been suggested (see Chapter 9) and it is likely that these receptors are involved in triggering cascades of kinases. Until these are precisely delineated a complete understanding will be lacking.

4.2 Non-receptor protein tyrosine kinases

In contrast to the receptor class, the non-receptor group of protein tyrosine kinases (see Table I) cannot be placed in any particular signal transduction pathway; neither input nor output signals have been defined. However there are a number of clues as to how and where they may fit and for the purposes of discussion the *c-src* encoded protein pp60 [c-src] (which is the protype for this class of kinases) will serve to illustrate this group of proteins.

Gene	Chromosome	Tissues
c-src	20q13.3	Ubiquitous
c-yes	18q21.3	Brain,placenta>liver, kidney,lung.
c-fyn	6q21	Brain>placenta, liver,kidney,lung.
c-fgr	1p36.1	B-cells
c-lck	1p32-p35	T-cells>>B-cells
c-hck	20q11-q12	Granulocytes>B-cells
c-lyn	8q13-qter	Placenta>liver>brain,kidney,lung

Table 1. Some members of the non-receptor tyrosine kinase family. The human chromosomal localization and tissue distribution of some members of this family are indicated.

pp60$^{c\text{-}src}$ is the cellular homologue of the *v-src* transforming protein (see below). The identification of this gene product (and other kinase genes in this class) preceded a functional characterization of the protein and much of what is known about this protein concerns structural aspects. pp60$^{c\text{-}src}$ is myristylated post-translationally on its amino-terminal glycine following removal of the initiator methionine residue. This myristylation is responsible for membrane association of the protein and its location there is almost certainly necessary for normal function of the protein. Thus the location of this kinase at the cytoplasmic face of the plasma membrane resembles the kinase domains of the receptor class.

Labelling studies and peptide mapping have revealed a number of phosphorylation sites on pp60$^{c\text{-}src}$ some of which appear to serve a regulatory function (figure 11). Of particular importance is the phosphorylation of tyrosine-527. Phosphorylation at this site (which is not an autophosphorylation event) leads to inhibition of the tyrosine kinase activity (see section 5); under most normal conditions this site is phosphorylated to high stoichiometry, implying that the kinase activity of pp60$^{c\text{-}src}$ is usually suppressed.

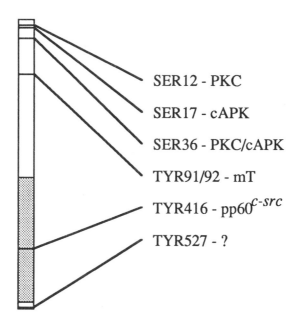

Fig 11. Phosphorylation sites on pp60$^{c\text{-}src}$. Some of the defined *in vitro* and *in vivo* phosphorylation sites of the *c-src* encoded protein are shown with the kinases responsible. The phosphorylation of tyrosine 91/92 occurs as a consequence of complex formation with middle T (mT) which is not itself a kinase. Tyrosine 416 is an autophosphorylation site. Phosphorylation of tyrosine 527 plays a critical role in the control of pp60$^{c\text{-}src}$ activity (see text), however the kinases(s) responsible for this phosphorylation has not been defined.

Comparison of the sequence of pp60 $^{c\text{-}src}$ and related non-receptor tyrosine kinases reveals that there is homology not only in the kinase domain *(src* homology 1 domain - SH1) but also in two regions outside this catalytic core; these are termed SH2 and SH3. It is these latter two regions which are also found in a series of otherwise unrelated polypeptides PLC-γ (SH2, SH3), GAP (SH2), spectrin (SH3), and the transforming protein *v-crk* (SH2, SH3). Based upon mutations in this domain in the *c-src* related *v-fps* tyrosine kinase, it would appear that the SH2 domain (and by inference the SH3 domain) is involved in interaction with another protein(s) and that this plays a permissive role in the function of the tyrosine kinase. The nature of the proteins that interact with these regulatory domains has yet to be determined.

With respect to the position in a control hierarchy that these non-receptor tyrosine kinases play, evidence suggests that "non-receptor" is a misnomer. The tyrosine kinase encoded by p56^{lck} is expressed in T cells; this kinase is associated with the transmembrane polypeptide CD4 (or CD8) in a complex that is dependent upon interaction of an amino terminal domain of the p56^{lck} protein with the short cytoplasmic tail of CD4 (or CD8). It is thought likely that activation of T-cells that express CD4 or CD8 occurs through interaction of these cell surface complexes perhaps with cell surface phosphotyrosine phosphatases (CD45); p56^{lck} has been shown to become activated in CD45 expressing T-cells (see Chapter 6). A simple model for the function of these tyrosine kinases is that they are in effect cytoplasmic subunits of receptor molecules that become activated through oligomerisation as shown in figure 12. Thus perhaps the main distinction between the receptor tyrosine kinases and the "non-receptor" group may simply be in their modes of activation.

5. Kinases in normal and aberrant control of cell division

It is not yet possible to trace cell surface events to specific aspects of the cellular control of division. However evidence for the operation of protein kinases in the processes of growth control comes not only from circumstantial evidence that a number of these kinases appear to become activated in response to mitogens, but also from various genetic approaches that highlight the importance of this class of proteins in normal growth and neoplasia.

5.1 Kinases in cell cycle control

The links between cell surface events and the mechanics[7] of the cell cycle remain obscure, however studies in the fission yeast *S. pombe* (which displays a cell cycle that is very similar to that observed in higher eukaryotes) have unearthed a series of regulatory proteins that are intimately involved in the control of progression through the cell cycle. For entry into mitosis, the network elucidated from genetic studies is summarized in figure 13. Interestingly *cdc2*, *niml* and *weel* genes all encode serine/threonine specific protein kinases, suggesting a cascade of controls responsible for timing entry into mitosis.

---

[7] For progression through the cell cycle certain criteria have to be met at various stages. For example, completion of S-phase must precede mitosis for two viable daughters to emerge. It is the operation of these inter-dependent events that is referred to as the 'mechanics'.

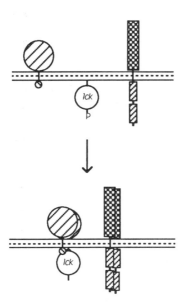

Fig 12. Model for the activation of p56^{lck}. Shown schematically are components of a putative signalling system and their membrane orientation. These are: CD4 (hatched), p56^{lck} (indicated as *lck*) and CD45 (check/hatch). The protein tyrosine kinase p56^{lck} is associated with the plasma membrane through a myristyl group attached to the N-terminus. On activation of T-cells with antibodies to CD4 the antigen is cross-linked and is found associated with *lck*. Activation of the p56^{lck} may require the dephosphorylation of the regulatory site at the C-terminus of this protein. This dephosphorylation may be catalysed by the cell surface tyrosine phosphatase CD45 (see chapter 6).

The powerful selection that can be applied to yeast has been used to identify a functional human homologue of cdc2 and current progress indicates that the control of the mammalian cell cycle, at the level of the mechanics of the operation, is well conserved relative to lower eukaryotes such as yeast. Thus kinases are critically involved in controlling the cell cycle. An obvious issue yet to be unravelled is precisely how extracellular events trigger/impinge on these endogenous control mechanisms?

5.2 Oncogenic protein kinases

There are a number of transforming genes that are protein kinases. These include the serine/threonine kinases encoded by *v-raf* and *v-mos* and a multitude of tyrosine kinases typified by *v-src* ("non"- receptor) and *v-erb B* (receptor related; see Chapter 3) gene products. The ability of these genes to cause transformation is due to their aberrant activation and it is because these critical regulatory proteins lie at key positions in the control hierarchy, that activation causes a pleiotropic growth response.

How are these genes activated? The *v-raf* gene as its prefix denotes is a form of the cellular *c-raf* gene that has been "picked up" by a retrovirus yielding *v-raf*[8]. In the acquisition of the *c-raf* gene by the retrovirus a substantial part of the regulatory domain of *c-raf* was deleted including a putative pseudosubstrate site and a cysteine-rich domain (figure 14).

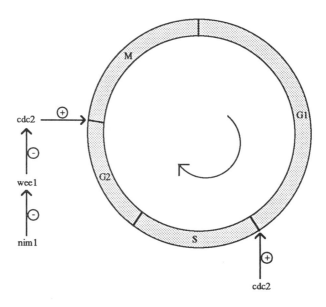

Fig 13. Protein kinases in cell cycle control. Three protein kinase genes have been identified that play regulatory roles in the control of the cell cycle in the yeast *S.pombe*. The gene *cdc2* operates at two critical points in the cell cycle, while genetic evidence suggests that the genes *nim1* and *wee1* function in some form of regulatory cascade impinging on the function of *cdc2* at the G2/M boundary. There is a mammalian homologue of *cdc2* which also appears to be important in cell cycle control; it is likely that there are a number of other elements in the control of cell division that are well conserved through evolution.

By analogy with PKC which is discussed above (section 3.1), this deletion would be expected to lead to expression of a constitutively active kinase. This prediction appears to be correct, since deletions of *c-raf* itself have been used to demonstrate the transforming potential of such a kinase mutant.

[8] Many transforming genes have been identified in various model systems through the ability of certain viruses - the RNA tumour viruses - to induce neoplasia. The genes responsible (oncogenes) are derived from cellular genes through recombination with the viral genome. Subsequent conversion to transforming genes appears to be effected through mutations in the new virally encoded gene. The nature of these mutations is discussed in the text.

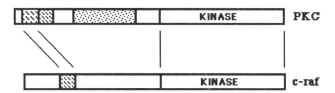

Fig 14. Structural similarity of PKC and the *c-raf* gene product. The two conserved regions in the regulatory domain of PKC are shown alongside the *c-raf* polypeptide. In both proteins the kinase domain is located in the C-terminal half of the polypeptide. In c-raf there is a cysteine rich region that resembles one of the repeat units present in PKC. It is inferred that this region of c-raf is part of an effector binding domain that is necessary for the regulation of c-raf. Deletion of the amino-terminus of c-raf including this region produces a transforming gene product (see text).

Effectively similar mechanisms operate in the deregulation of the non-receptor tyrosine kinases. The most critical control of this group of proteins is through phosphorylation of a C-terminal tyrosine (equivalent to tyrosine-527 in $pp60^{c-src}$). These C-terminal sequences are deleted in retrovirally acquired transforming genes of this class. In fact point mutation of tyrosine -527 in the *c-src* polypeptide to phenylalanine leads to the generation of a gene with transforming potential.

There is also a second interesting alternative mechanism that leads to $pp60^{c-src}$ activation and this involves the middle T gene (mT) of the DNA tumour virus polyoma. It has been shown that mT will associate in a complex with $pp60^{c-src}$ leading to the activation of the tyrosine kinase activity. The activation correlates with a low stoichiometry of phosphorylation of tyrosine-527, although it is not clear whether mT binding blocks phosphorylation or activates dephosphorylation. Nevertheless this means of activation occurs independent of any alteration in the *c-src* gene.

In the context of oncogenicity, the role of PKC as the major target for the action of phorbol esters and other tumour promoters of this class should be mentioned. The promotion induced by these agents is a complex phenomenon that involves both a stable change in the phenotype of the target (initiated) cell and also the ability of the promoters to stimulate hyperplasia of these cells with the consequent outgrowth of clones of cells with a growth advantage. Thus in the model for multistage carcinogenesis it can be envisaged that alterations to PKC activity may contribute to neoplasia. An indication that mutations in PKC can effect transformation is afforded by the observation that one particular uv-transformed fibroblast cell line expresses a PKC gene that carries four point mutations; the transfection of this mutant PKC into the parental fibroblast cell line induces a transformed phenotype. The consequence of these mutations on the behaviour of the PKC has yet to be defined, but one would expect that they induce a deregulated kinase activity as described above for other transforming kinase genes.

6

Protein phosphatases

1. Regulation of protein phosphatases adds sensitivity and flexibility to processes controlled by phosphorylation

As discussed in the previous chapter, phosphorylation-dephosphorylation of proteins on serine, threonine and tyrosine residues represents a major cellular control mechanism for the rapid and reversible regulation of cellular functions. The phosphorylation state of a protein is the result of a dynamic process, with the distribution between the phospho- and dephospho- forms of an interconvertible enzyme being governed by the relative activities of both protein kinases and protein phosphatases. In order for a signal transduction pathway to function effectively, it is essential that either the protein kinase, or the protein phosphatase, or both, are regulated directly or indirectly by the stimulus in question. Under appropriate conditions even a monocyclic interconvertible enzyme cascade system can generate an enormous sensitivity. Such a system shows an increased sensitivity if the effector acts coordinately to activate one element and inhibit the opposing element. This condition is fulfilled in an analogous situation where kinases and phosphatases are interlinked in a second cascade wherein the same phosphatase not only dephosphorylates the primary substrate but also the kinase, leading to inactivation of the kinase (see figure 1). In this situation a small stimulation of the phosphatase activity by an effector will automatically be amplified (because of its catalytic nature) as a dephosphorylation of the kinase. Such complex systems are known to operate in the control of enzymes that regulate intermediary metabolism.

Coordination of the regulation of different metabolic pathways and other cellular events is possible because most of the protein kinases and phosphatases are multifunctional, influencing several different processes at the same time. The number of identified protein kinases is increasing and explains the assertion that the mammalian genome may encode as many as "a thousand and one kinases" (see Chapter 5). In contrast to the overwhelming number of kinases, only a relatively small number of multifunctional phosphatases exist to control the dephosphorylation of regulatory proteins. It is interesting to note that systematic searches for new classes of cellular phosphatases using substrates phosphorylated by specific kinases has revealed only "known" phosphatases. It has been only with the use of molecular cloning techniques that homologous sequences (and by inference gene products) have now been detected. The characteristics of these protein phosphatases and their role in signal transduction and growth control is discussed below.

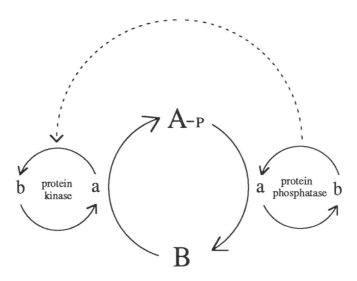

Fig 1. A protein kinase-phosphatase cascade system. The sensitivity of the system for the interconversion of a target protein from its active phosphorylated state (A-P) to its inactive dephosphorylated state (B) is greatly enhanced by the inactivation of the protein kinase (a-b conversion) by the active protein phosphatase (a). Thus activation of the protein phosphatase not only increases the conversion of A-P to B but also inhibits the rephosphorylation of B through inactivation of the kinase.

2. The protein phosphatase gene families

Two major classes of protein phosphatases exist, (a) those specific for phosphoserine/ threonine residues, and (b) those specific for phosphotyrosine. In each class there are several different subtypes that can be distinguished on the basis of structure and regulation.

Four different types of serine/threonine phosphatases can be distinguished (Table 1): the polycation-stimulated (PCS or type 2A) protein phosphatases, the ATP,Mg-dependent phosphatases (AMD or type 1), the Ca^{2+}-calmodulin stimulated phosphatase (calcineurin or type 2B) and the Mg^{2+}-dependent phosphatase (type 2C). The other major class of protein phosphatases, specific for tyrosine residues, can be broadly classed into two distinct groups: low molecular weight, cytoplasmic enzymes, and high molecular weight, membrane associated receptor phosphatases.

Protein phosphatases can be regulated by several different types of control mechanisms: by allosteric interactions with small molecules, through dissociation/ association of subunits or aggregation with other proteins, through differential cellular localization and by post-translational modification. Since the substrates for phosphatases are also proteins, complex regulatory mechanisms can operate at the substrate level as well. Not only specific ligand binding but also covalent modification of the substrate can alter the protein phosphatase activity. Most phosphorylatable proteins contain multiple phosphorylation sites, and complex site-site interactions can influence the rate of phosphorylation or dephosphorylation of another site. The combined action of two phosphatases or a phosphatase and a kinase may be necessary in order to obtain dephosphorylation of a crucial site influencing the properties of the target protein. Multiple phosphorylation sites, governed by different combinations of kinases and phosphatases, represent the mechanism by which enzymes or other regulatory proteins can respond to integrate several physiological stimuli and by which stimuli can influence each other (discussed further in Chapter 8).

Class	Subtype and mode of regulation	Abbreviation	Phosphosubstrate	Numbering system
I	Serine/threonine specific phosphatases			
	ATP,Mg-dependent	AMD	P-Ser; P-Thr	1
	Polycation-stimulated	PCS	P-Ser; P-Thr; (P-Tyr)	2A
	Ca^{2+}/calmodulin-dependent	Calcineurin	P-Ser; P-Thr; (P-Tyr)	2B
	Mg^{2+} dependent		P-Ser; P-Thr	2C
II	Phosphotyrosine phosphatases			
	Cytoplasmic (soluble)	PTPase	P-Tyr	3
	Receptor-like	RPTPase	P-Tyr	3

Table 1. Classification of protein phosphatases.

2.1 Serine/threonine protein phosphatases

The serine/threonine specific protein phosphatases are a diverse group of enzymes that are regulated by different mechanisms and have different substrate specificities. Molecular cloning of the catalytic subunits has revealed an interesting feature. The deduced amino acid sequences of the PCS protein phosphatases, the AMD phosphatase and calcineurin are related, suggesting that they have gained distinctive regulatory properties during evolution from a common ancestral gene (figure 2). The other major serine/threonine protein phosphatase, the Mg^{2+}-dependent phosphatase, has also been cloned. Interestingly it does not show any sequence homologies to the other catalytic subunits suggesting that it has evolved from a different ancestral gene.

The PCS-protein phosphatases (protein phosphatase 2A)

At least four distinct oligomeric PCS phosphatases have been characterized from many different species and tissues (Table 2). The different PCS phosphatases not only differ in subunit structure and molecular weight, but also in other physical and enzymatic properties. A common structural feature of all these enzymes is a catalytic subunit (PCS$_C$) of molecular mass 36,000 and a putative regulatory subunit of molecular mass 65,000 (PR65). In addition the PCS$_{H1}$ phosphatase contains a 55,000 and the PCS$_M$ phosphatase a 72,000 subunit. Although the PCS$_{H2}$ and PCS$_L$ phosphatases display a similar subunit structure they differ in substrate specificity. Moreover, the PCS$_{H2}$ is apparently derived from the PCS$_{H1}$ phosphatase by loss of the 55,000 subunit. The similar subunit structure of the PCS$_{H2}$ and

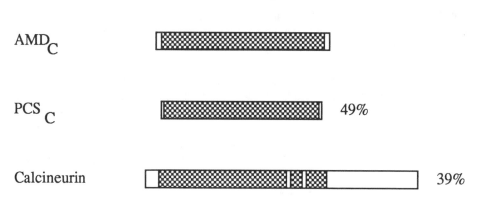

Fig 2. A comparison of the related serine/threonine-specific protein phosphatases. The chequered regions indicate those conserved between the different phosphatases; the percent identity is given with reference to the AMD_c protein. The unique C-terminal domain of calcineurin includes the calmodulin binding site.

PCS_L phosphatases but different enzymatic properties suggests the existence of distinct but related genes for the regulatory subunits or perhaps post-translational modification of one or other subunit.

The molecular weight of the PCS_{H1} phosphatase in cell extracts is substantially higher than the molecular weight of the purified enzyme, therefore it is anticipated that this phosphatase interacts with other cellular components or structures which are lost during the purification procedure. The PCS phosphatases are considered as cytosolic enzymes, however membrane bound, chromatin bound and nucleoplasmic locations have also been observed.

The activity of the PCS phosphatases can be directly stimulated *in vitro* with polycations such as protamine, polylysine or histone H1 (lysine rich). Polycation stimulation can be used to distinguish isotypes of PCS-phosphatases; however polycation stimulation is dependent on the substrate used. The PCS-phosphatases are not affected by the heat stable inhibitors, inhibitor-1 and the modulator subunit (also termed inhibitor-2) which are specific inhibitors of the AMD phosphatase (see below). The PCS phosphatases show a broad substrate specificity which is overlapping but distinct from the specificity of the

Type	Subunits			
	C	PR65	PR55	P75
PCS_{H1}	+	+	+	
PCS_{H2}	+	+		
PCS_M	+	+		+
PCS_L	+	+		

Table 2: Subunit structure of purified PCS phosphatases.

AMD phosphatase. As with the protein kinases, use has been made of synthetic peptides in the determination of structural requirements for phosphatase substrates. Such an analysis is summarized in Table 3.

The functional role of the PCS phosphatase regulatory (PR) subunits, i.e. non-catalytic subunits, is not yet fully understood, but reconstitution experiments of the catalytic subunit with the PR subunits have demonstrated changes in activity and substrate specificity. The PR65 derived from a pig heart PCS phosphatase preparation increases dramatically the phosphatase activity of the catalytic subunit with an histone H1 substrate. By contrast the PR55, which binds directly to the PR65, suppresses the dephosphorylation of phosphorylase

1) Preferential dephosphorylation of phosphothreonyl peptides[a]

Phosphopeptides	Dephosphorylation rate (pmol/min/ml)	
	Phosphorylase a	Peptide
RRA(Tp)VA	150	1,484
RRA(Sp)VA	150	32
RRREEE(Tp)EEEAA	150	1,417
RRREEE(Sp)EEEAA	150	126

2) C-terminal proline residues as a strong negative determinant[b]

RRA(Tp)VA	500	4,760
RRP(Tp)VA	500	1,870
RRA(Tp)PA	500	7
RRP(Tp)PA	500	7

3) N-terminal cluster of basic residues in -3 to -6 position as positive determinants for dephosphorylation of phosphoseryl peptides[b].

RRA(Sp)VA	300	53
RRRR(Sp)VA	300	64
RRRA(Sp)VA	300	207
RRRRA(Sp)VA	300	407
RRRRAA(Sp)VA	300	913

Table 3. Analysis of primary sequence determinants for the PCS phosphatases. Some of the recognition determinants in the primary structure of phosphorylation sites have been investigated using model peptide substrates. This analysis has revealed that phosphothreonyl containing peptides are better substrates than their phosphoseryl counterparts and that a cluster of basic amino acid in the -3 to -6 position is a positive determinant for dephosphorylation of phosphoseryl peptides. The PCS phosphatases also efficiently dephosphorylate phosphopeptides with acidic clusters downstream of the phosphorylated residue, a feature required for phosphorylation by casein kinase II. Interestingly, a proline residue C-terminal to a phosphothreonine is a strong negative determinant for dephosphorylation by the PCS phosphatases. Such a -Thr-Pro- phosphorylation site would, therefore, form the structural basis for protection from dephosphorylation by the PCS phosphatases.
[a] The type of phosphatase used in these experiments was the PCS_M phosphatase.
[b] The type of phosphatase used in these experiments was the PCS_H phosphatase.

and glycogen synthase with marginal effects on the dephosphorylation of histone H1. Similarly reassociation of the PR65 subunit isolated from reticulocytes increases the activity of the PCS$_c$ subunit from the same tissue, when the initiation factor eIF-2α is used as substrate, whereas the same polypeptide inhibits the dephosphorylation of 40S ribosomal subunits. These observations illustrate the means by which the PCS phosphatases show a diversity of function.

Molecular cloning techniques have revealed the existence of two very highly conserved and homologous cDNA sequences for the PCS$_c$ subunit; these are termed α and β. The encoded proteins are the products of distinct genes that differ mainly in their promoters and 3' noncoding region; only minor differences exist within the coding region. There are also two forms of the PR65 subunit, as shown by cDNA cloning techniques (termed PR65 α and β). These are apparently products of separate genes, and their predicted amino acid sequences are 85% identical. The α and β isotypes are found to be made up of 15 imperfect repeat units consisting of 39 amino acids and this repeating structure is conserved between isotypes and species from man to *Drosophila* suggesting that this unique structure plays some as yet undefined critical role.

The physiological significance of the existence of the multiple PCS$_c$ and PR65 isotypes is not yet known. It is still not clear whether the isotypes of PCS$_c$ and PR65 show selective associations or are randomly distributed between the different holoenzyme forms. The availability of both cDNAs for the PCS$_c$ and PR65 enables investigation of these basic questions.

The ATP,Mg-dependent (AMD) phosphatase (phosphatase 1)

Several different forms of the AMD phosphatase have been purified. The most extensively studied form is a catalytically inactive heterodimer consisting of a catalytic subunit (AMD$_c$ also abbreviated F$_c$) of molecular mass 38,000 and a heat stable modulator subunit (also known as inhibitor-2) of 23,000. This form of the AMD phosphatase can be activated by transient phosphorylation of the modulator subunit by a protein kinase (see below). The AMD phosphatase can also be isolated as a fully active free catalytic subunit; however, the vast majority (up to 95%) is present in the cell in an inhibited state, repressed by one of several mechanisms.

Two full-length clones for the AMD$_c$ subunit have been isolated from a rabbit skeletal muscle cDNA library; these are AMD$_c$ α and AMD$_c$ β. These cDNAs appear to be derived from the same gene by alternative splicing since the predicted amino acid sequences are identical with the exception of the 33 N-terminal amino acids (AMD$_c$α), which are replaced by a 14 amino acid long sequence in AMD$_c$β. However, the existence of multiple genes for the AMD$_c$ subunit remains a possibility since studies on the AMD phosphatase from *Drosophila* indicate three distinct genetic loci.

The AMD phosphatase can be regulated in several ways. The activity of the catalytic subunit can be changed through the action of a specific protein kinase (known as protein kinase F_A or as glycogen synthase kinase 3). The mechanism of this activation is known in some detail: a transient phosphorylation of the modulator (M) subunit (on Thr-72) constitutes the first step, followed by a conformational change in the catalytic subunit resulting in the production of an active protein phosphatase. Auto-dephosphorylation appears to be essential in the activation process (figure 3).

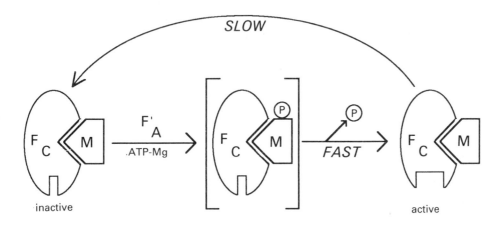

Fig 3. The activation cycle for the AMD phosphatase. The kinase F_A phosphorylates the modulator subunit (M) which interacts with the catalytic subunit of the AMD phosphatase F_C. This phosphorylated state autocatalytically converts to a dephosphorylated active species of $F_C M$. This active form slowly converts to the inactive species through a conformational change.

Two other protein kinases, casein kinase I (CKI) and casein kinase II (CKII) are also implicated in the regulation of the AMD phosphatase. Interestingly phosphorylation of the modulator subunit by CKII does not directly influence activity, but potentiates the F_A-mediated activation. The activated AMD phosphatase does not autodephosphorylate the CKII site, but PCS_H phosphatase will. This interplay of phosphorylation represents an example of permissive control acting through $CKII/PCS_H$ on the AMD phosphatase. In contrast to the above the CKI phosphorylation of modulator blocks subsequent activation by protein kinase F_A. This represents a simple inhibitory input signal.

The activity of the AMD phosphatase can be further regulated by two additional proteins, inhibitor-1 and its antagonist, the deinhibitor protein, both are regulated in a coordinated way by the cAMP-dependent protein kinase and the PCS phosphatase. Inhibitor-1 is a heat stable protein of molecular mass 21,000. Inhibitor-1 only functions as an inhibitor after phosphorylation of Thr-35 by the cAMP-dependent protein kinase. In brain an additional isoform of inhibitor-1 exists; this is termed "dopamine and cAMP-regulated phosphoprotein" DARPP.

Whereas the purified AMD_c subunit is very sensitive to the phospho form of inhibitor-1 (K_i in the nanomolar range), partially purified forms of the AMD phosphatase are rather resistant to its inhibition and require 1000 fold higher concentrations for inhibition. This observation led to the isolation of a heat stable protein of molecular mass ~9,000 - the deinhibitor protein - that is only active in its dephospho form. The deinhibitor protein antagonizes the effect of inhibitor-1 and stabilizes the AMD phosphatase in its active or partially activated state preventing the conversion into the inactive conformation. Moreover it releases the inhibitory constraints imposed by other inhibitory proteins. In contrast to inhibitor-1, the deinhibitor is inactivated by the cAMP-dependent protein kinase. Thus regulation of AMD phosphatase activity by the cAMP-dependent protein kinase is coordinated through both activation of an inhibitor and inhibition of a deinhibitor! As discussed in section 1 such concerted regulation provides an exquisitively sensitive control mechanism. A schematic diagram summarizing AMD phosphatase regulation is presented in figure 4.

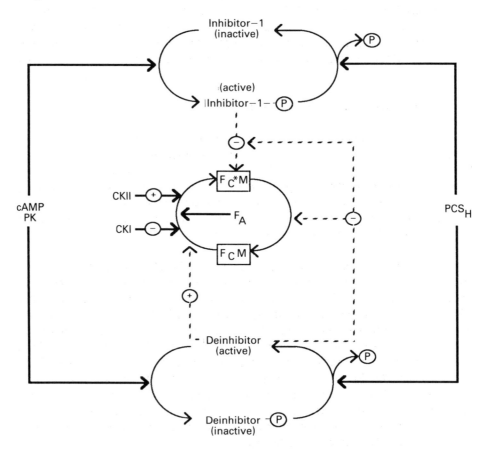

Fig 4. Regulation of the AMD phosphatase. The AMD_c-modulator complex (F_cM) exists in active (F_c*M) and inactive conformations, interconverted through the action of the kinase F_A (see also figure 3). The activity and stability of F_c*M is also determined by inhibitor-1 and the deinhibitor; both of these regulators exist in interconvertible forms. The action of various kinases are shown in bold arrows, as are the sites of action of the PCS_H phosphatase. The regulatory influences are shown in dotted lines.

The AMD_c subunit also associates with other proteins which can determine its cellular location, and/or substrate specificity. A phosphatase isolated from skeletal muscle, termed protein phosphatase-1G, consists of a regulatory glycogen binding subunit (G-subunit) and the catalytic subunit of the AMD phosphatase. Phosphorylation of the G-subunit by the cAMP-dependent protein kinase leads to dissociation of AMD_c, while the G-subunit remains bound to glycogen.

Other molecular species of the AMD-phosphatase have been described. In liver a minimal three-subunit structure has been proposed for the glycogen bound enzyme. In rabbit skeletal muscle, a large quantity of the AMD phosphatase is specifically bound to myosin, and a phosphatase 'receptor' has been described to explain the specific binding of the AMD_c subunit to microsomal membranes. In the isolated nuclei of *Xenopus* oocytes, a similar form of the AMD phosphatase has been identified and shown to be present in the nucleoplasm, whereas the AMD phosphatase was also associated with the nuclear particulate fraction. Whether, and how, the same catalytic subunit can shuttle between these different cellular locations is unknown. However it remains an attractive hypothesis that targeting proteins exist to sequester the phosphatase in particular compartments of the cell and that by doing so they serve to introduce specificity into the responses elicited by phosphatase (in)activation.

Ca^{2+}-Calmodulin stimulated phosphatase (calcineurin, phosphatase type 2B)

Calcineurin is a heterodimer consisting of A (61,000) and B (19,000) polypeptides. The A subunit can be divided into three domains: the catalytic site, a high affinity (nM) calmodulin binding site and a binding site for the B subunit. The primary structure of the catalytic A subunit reveals significant homology with the AMD_c and PCS_c phosphatase (see figure 2); in addition the calmodulin binding site shows significant homology to other calmodulin-binding proteins. The B subunit is a Ca^{2+} binding protein, possessing four Ca^{2+} binding sites, and an N-terminal glycine residue that is myristylated.

Calcineurin was originally discovered as a major calmodulin binding protein from bovine brain and isolated as an inhibitor of the Ca^{2+}/calmodulin-dependent cyclic nucleotide phosphodiesterase. Later it was identified as a Ca^{2+}/calmodulin stimulated phosphatase with a rather narrow substrate specificity. The high affinity calmodulin binding and phosphatase activation is dependent on the presence of Ca^{2+}, although Ca^{2+} in the absence of calmodulin is also slightly stimulatory (see figure 5). This phosphatase has been found in all tissues examined and is present even in lower eukaryotes. In higher organisms the brain contains 10 to 20 times as much calcineurin as other tissues and levels of as high as 1200 mg/kg wet brain tissue have been reported. Multiple forms of calcineurin have been detected by molecular cloning. Interestingly the deduced amino acid sequences reveal a tract of polyproline (11 residues) in the aminoterminal domain; this is thought to be involved in the Ca^{2+}/calmodulin-dependent activation.

Phosphorylation by the Ca^{2+}/CaM-dependent protein kinase II or protein kinase C results in partial inactivation of calcineurin due to an increase in K_m; this phosphorylation is blocked by prior binding of Ca^{2+}/CaM. Whereas auto-dephosphorylation can be very slow, calcineurin is rapidly dephosphorylated by what appears to be a brain PCS phosphatase.

Several phosphoproteins present in the brain are potential substrates for calcineurin, including the AMD phosphatase inhibitors, inhibitor-1, DARPP and G-substrate, the microtubule associated proteins MAP-2 and tau, myelin basic protein and the type II regulatory subunit of the cAMP-dependent protein kinase. There is evidence for a role of calcineurin in the regulation of the voltage-activated Ca^{2+} channels, essential for nerve cell function.

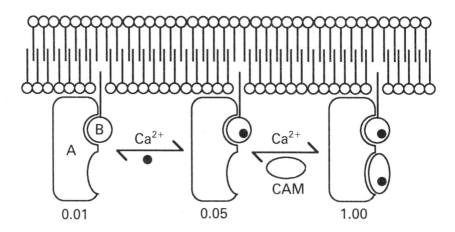

Fig 5. Subunit structure and activation of calcineurin. The A-B heterodimer is shown associated with the membrane through the myristyl group which is covalently attached to the amino terminus of the regulatory B subunit. In the absence of calmodulin (CAM), calcium binding to the B subunit causes partial activation. In the presence of both calmodulin and calcium, full activation is achieved. Approximate relative activities of the different species are indicated underneath.

Calcineurin is also reported to dephosphorylate some phosphotyrosyl residues in proteins. However a functional role for calcineurin as a phosphotyrosyl phosphatase has been questioned on the basis that this activity is highly dependent on Ni^{2+} and Mn^{2+} ions.

The Mg^{2+}-dependent phosphatase (phosphatase 2C)

The Mg^{2+}-dependent phosphatase is a monomeric enzyme of molecular mass 43,000 that has been purified as a myosin (P) light chain phosphatase, as a glycogen synthase phosphatase and as a 6-phosphofructo-1-kinase phosphatase. It represents a major activity towards the latter enzyme as well as towards pyruvate kinase, fructose-1,6-bisphosphatase and 6-phosphofructo-2-kinase, and therefore might play a major role in the regulation of gluconeogenesis; by contrast its function as a glycogen synthase phosphatase has been questioned on a quantitative basis.

Two forms of the Mg^{2+}-dependent phosphatase have been isolated, and based upon structural studies these probably represent the products of different genes. A full-length cDNA for the Mg^{2+}-dependent phosphatase has been isolated from a kidney cDNA library. Interestingly the predicted protein sequence shows no relationship to the AMD phosphatase, PCS phosphatase or calcineurin. Studies involving the expression of the cDNA in mammalian cells should lead to an understanding of its role *in vivo*.

2.2 Phosphotyrosyl specific phosphatases (PTPases)

Phosphorylation of cellular proteins on tyrosyl residues has been implicated in the regulation of cell proliferation, differentiation and transformation. Recent advances in the study of PTPases suggest that these may be primary targets involved in cell cycle regulation. Furthermore, the discovery that a major PTPase isolated from human placenta (see below) shows an unexpected sequence homology with the intracellular domain of the leukocyte common antigen CD45, suggests that this protein could be a phosphatase regulated via its extracellular domain by an as yet unidentified ligand.

Cytoplasmic PTPases

The dearth of proteins phosphorylated on tyrosine residues *in vivo* has significantly hampered the identification of PTPases. Much work has centred on the use of the synthetic peptides or artificially produced protein substrates. This latter approach has been most successful and led to the identification of multiple activities from tissues. One of these, PTPase 1B, has been purified to homogeneity from human placental tissue cytosol and its sequence directly determined by conventional microsequencing techniques. This advancement has led to the isolation of cDNAs encoding several different forms of PTPase. The determined and predicted amino acid sequences of the different forms of PTPase do not show any similarity with the serine/threonine phosphatases.

The specific activity of the purified PTPase 1B enzyme is high and it is anticipated that *in vivo* the regulation of this activity may be through inhibitory influences. Indeed in brain two different inhibitory proteins of PTPases have been described. In addition, in *Xenopus* oocytes, inactivation of a PTPase 1B-like activity by an ATP,Mg dependent process has been observed.

Clues to the function of the PTPases has come in part from their artificial introduction into physiological settings. For example, microinjection of PTPase 1B into *Xenopus* oocytes is able to antagonize the action of insulin which would normally induce maturation and can acutely abolish the activation of short-term insulin responses.

Other experiments have shown that overexpression of the T cell PTPase through introduction of the cDNA, can lead to disruption of nuclear regulation resulting in multinucleated cells. This phenotyic change in T cells, in some respects resembles "wee" mutants in yeast where the cells appear to become advanced in mitosis and divide inappropriately before the cell has reached double its size (so producing "wee" daughter cells).

These examples indicate a role for PTPases in controlling oocyte maturation and mitosis; the extent to which this is an active as opposed to a passive role has yet to be determined.

The receptor like phosphotyrosyl phosphatases (RPTPase)

The sequence of the purified placental PTPase led to the discovery that receptor-like molecules could also function as phosphatases. It was noted that the sequences of the placental PTPase shared some homology to the two tandem (internally homologous) C-terminal domains of the leukocyte common antigen (CD45). The function of this well known family of membrane spanning molecules was not known prior to this discovery, but recent experiments demonstrate that CD45 has an intrinsic PTPase activity.

The structural organization of CD45 resembles that of the growth factor receptor protein tyrosine kinases, consisting of a large cysteine rich, glycosylated external domain, a membrane spanning region and a large C-terminal cytoplasmic region with catalytic function; these PTPases have thus been termed "receptor PTPases" (RPTPases).

Using a CD45 cDNA probe another putative RPTPase, termed LAR (leukocyte common antigen related protein) was isolated from a placental cDNA library. Its predicted structure has the same basic organization as CD45, and the mRNA is expressed in many tissues and cell types. Using oligonucleotide probes, corresponding to a conserved, presumed PTPase catalytic domain, two PTPase genes from *Drosophila* have also been cloned; these are termed DLAR and DPTP. The predicted structures contain tandem catalytic domains, a membrane spanning domain and large external domains again suggesting that they are receptor PTPases. The structures of these various PTPases are summarised in figure 6.

Although the cytoplasmic domains of LAR and CD45 are similar, their external domains are completely unrelated. The large extracellular LAR segment shows similarity to the neural-cell adhesion molecule (N-CAM), a member of the immunoglobulin superfamily. These adhesion molecules are known to play a role in cell-cell interactions, controlling morphogenesis and tissue development. Therefore it has been suggested that the LAR family of receptor-like PTPases might function in cell-cell interactions. For example,by antagoniszing the action of growth factors through receptor tyrosine kinases, LAR and related RPTPases may be responsible for the cessation of growth associated with cell-cell interaction (contact inhibition).

The PTPase activity associated with PCS phosphatases

Recent observations suggest that PTPase activity can be associated with the PCS_L phosphatase. A protein factor, termed PTPA (for phosphotyrosyl phosphatase activator) can specifically activate the PTPase activity of the PCS_L phosphatase in a Mg-ATP dependent reaction (no change in the seryl/threonyl phosphatase activity is seen). The activation by PTPA is specific for the two subunit PCS phosphatases (PCS_L and PCS_{H2}) suggesting that the function of the PR55 subunit in PCS_{H1} and the PR72 subunit in PCS_M prevent PTPA mediated activation. The mechanism of catalysis is clearly different from the previous group of PTPases in that the PTPA activated PTPase is completely dependent on metal ions. This might suggest that the substrate specificity of both groups of PTPase will be different.

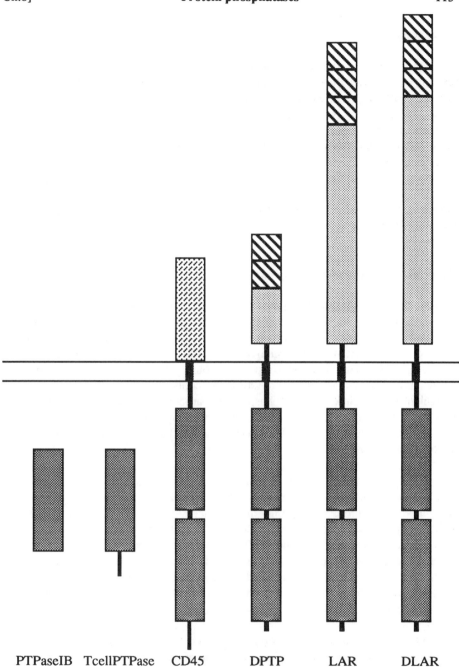

Fig 6. Structural similarities between soluble PTPases and receptor-like PTPases. The homology between the placental PTPase IB and the Tcell PTPase is 74%; the homology with the receptor-like PTPases is ~30%. The extracellular domains of DPTP, LAR and DLAR show sequence similarity to the neural adhesion molecule N-CAM (hatched boxes).

3. Role of phosphatases in signalling pathways

It should be clear from the previous sections that phosphatases are not just constitutively active but utilize many different regulatory mechanisms to modulate phosphatase activity. However, in only a few cases has it been demonstrated conclusively that regulation of a phosphatase(s) is the cause of the changes in enzymatic activity of target proteins. In the past, research on regulation by phosphorylation/dephosphorylation has focused on intermediary metabolism and the signalling pathways that use cyclic AMP, phosphoinositides or Ca^{2+} as second messengers. Recent progress in the unravelling of central biological puzzles such as the cell cycle, has revealed that here as well, the phosphorylation state of key proteins is crucial in the decision making of the cellular machinery. The following section illustrates examples of the role of protein phosphatases in signal transduction and the control of cell growth.

3.1 Hormonal regulation of phosphatases

The most direct example of second messenger regulation of a phosphatase is afforded by calcineurin. The dependence of this enzyme on Ca^{2+} (and calmodulin) places it within a large family of regulatory proteins which includes a number of protein kinases as discussed in the previous chapter. While activation of calcineurin is a likely consequence of elevated cytosolic Ca^{2+}, the restricted substrate specificity observed for this phosphatase limits the role for this enzyme. In fact many of the substrates defined for calcineurin are involved in the regulation of protein kinases and phosphatases particularly those phosphorylated by the cAMP-dependent protein kinases. Thus calcineurin may operate in part to modulate parallel pathways (see Chapter 8). The extent to which this role influences growth control processes has yet to be defined.

Hormonal regulation of the AMD phosphatase directly or through its regulatory subunits has been demonstrated *in vivo* in several systems. For example, the phosphorylation state of inhibitor-1 is controlled, in part, by intracellular cAMP levels; following adrenaline treatment of rabbits, the phosphorylation state of the inhibitor in skeletal muscle increases as a consequence of activation of the cAMP-dependent protein kinase. Conversely, the β-adrenergic antagonist propanolol can promote a decrease in the phosphorylation state of inhibitor-1. Insulin can also promote a decrease in the basal state of phosphorylation of inhibitor-1 in muscle and fat pads, but this effect seems to be secondary to the control exerted by cAMP.

Alterations in the phosphorylation of other regulatory proteins is also seen. In rodent fat cells, a rapid increase in the phosphoserine content of the modulator subunit is observed after insulin treatment. It is anticipated that this reflects an increase in AMD phosphatase activity, presumably through protein kinase F_A, although how this signal pathway fits together is unknown. An indication is perhaps provided by the observation that in human platelets insulin promotes the translocation of the protein kinase F_A from the membrane to the cytosolic fraction; this may be necessary for subsequent phosphorylation of the modulator subunit.

The adrenaline induced phosphorylation of the G subunit in skeletal muscle (responsible for targeting the AMD catalytic subunit to the glycogen particle), leads to the release of the catalytic subunit into the cytosol. Thus both the activity and localisation of the AMD phosphatase is subject to acute regulation through the action of hormones. In part this regulation is effected through the control of cAMP-dependent protein kinase, i.e. via a cAMP second messenger, however as indicated above for insulin, other factors also operate on this phosphatase by as yet unidentified mechanisms.

3.2 Phosphatases and regulation of the cell cycle

Three major control points have been identified in the cell cycle. Cells in rest (G0) or after each mitotic division have to pass a point early in G1 that commits the cell to initiating DNA synthesis (the S phase). Another control point is late in G2 (prophase), the cell phase separating the S phase from a next M phase, and finally there is exit from mitosis which is also a highly regulated control point. Based upon genetic analysis in yeast it is now believed that each of these transitions is regulated by controlling the activity of a single protein kinase, pp34(cdc2), the product of the *cdc2* gene. The amount of cdc2 protein remains relatively constant during the cell cycle and regulation is mediated by a combination of phosphorylation-dephosphorylation and interaction with other proteins. The best studied transition is the G2/M transition (initiation of mitosis). Mitotic cells contain an activity known as maturation-promoting factor (MPF) that can induce interphase cells to progress into mitosis. MPF has been purified and there is now general agreement that it is composed of two components: the cdc2 kinase subunit and the protein cyclin B. The histone H1 kinase activity of cdc2 cycles during the cell cycle and parallels changes in MPF activity. Underlying these changes in activity are events controlled in part through the action of protein phosphatases as described below.

The phosphorylation state of cdc2 is critical in regulating function; twelve different phosphorylation states of cdc2 have been described, and the phosphorylation sites include not only serine and threonine but also tyrosine residues. Dephosphorylation of tyrosine and threonine residues occurs as cells enter mitosis and this is indeed necessary for the G2 to M transition. The phosphatases responsible for these dephosphorylations are not yet identified but the complexity of the phenomenon suggests the operation of sophisticated regulation for these phosphatase(s).

The necessity for the dephosphorylation of cdc2 in the G2/M transition provides strong circumstantial evidence for the role of phosphatases in cell cycle control. More direct evidence, in this case for control of mitotic progression, has come from genetic studies. Recent investigations using three different fungi have revealed that the AMD phosphatase plays an important role in mitosis. Temperature sensitive mutants of *Aspergillus nidulans*, which are blocked in mitosis (*Bim* mutants) can be rescued by a wild type gene whose predicted amino acid sequence has a very high degree of homology to the mammalian AMD phosphatase catalytic subunit.

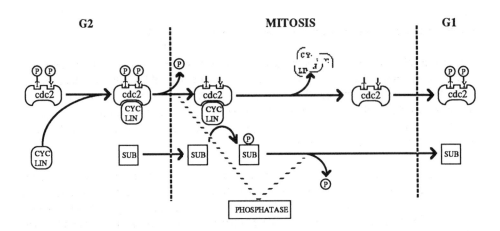

Fig 7. Phosphorylation of pp34 (cdc2) and M-phase control. Two critical events in the timing of the G2-M transition are the formation of a cdc2-cyclin complex and the dephosphorylation of the cdc2. The dephosphorylation involves both a threonine (T) and a tyrosine (Y) site as indicated. This dephosphorylated complex actively phosphorylates substrates (SUB) critical for the process of mitosis. For exit from mitosis, cyclin must be degraded and further undefined dephosphorylation events must occur (see text). In G1, cdc2 is returned to its inactive state in preparation to receive cues to trigger a further G2/M transition.

Genetic techniques have also revealed that in the fission yeast *Schizosaccharomyces pombe* two different genes could complement cold sensitive mutants defective in sister chromatid disjoining (*dis* mutants); although located on two different chromosomes, both genes *dis 1* and *dis 2* encode proteins (p37 and p39) with very similar predicted amino acid sequences, highly homologous (75-90% identical) to the AMD phosphatase catalytic subunit (AMD_c). Moreover the two gene products have overlapping functions since disruptants are lethal only when both genes are disrupted.

Analysis of a third group of mutants from *S.pombe* has demonstrated that the AMD_c phosphatase acts also in the pathway regulating the initiation of mitosis. The strategy followed in this instance was complex and underscores the power of genetic screening. In *S.pombe*, *wee1* mutations lead to small cells ("wee"); this mutant phenotype can be reversed by mutations in a second gene *cdc25*. Thus *cdc25* mutants can suppress the wee phenotype. In a cell expressing both mutant *wee1* and *cdc25*, it is possible to search for genes that overcome the suppression of the *wee1* phenotype by *cdc25*. Such a gene has been identified and termed *bws1* (bypass of wee suppression). This gene turns out to be identical to *dis 2* (see above). Thus yeast AMD phosphatase plays a critical role in the G2/M transition.

From these studies it is clear that the AMD phosphatase is important in both the initiation of and exit from mitosis. Although it was observed that the gene acts pleiotropically it remains a mystery how a mutation in the AMD phosphatase that is considered to be a multifunctional phosphatase intervening in many cellular processes, can have such a well defined, almost singular effect. Defining the substrates specifically involved in generating this phenotype will certainly help to solve this problem.

4. Phosphatases and the aberrant growth of mammalian cells

Two recent observations have dramatically changed our perception of serine/threonine protein phosphatases and their role in cellular growth control. Firstly a potent tumour promoter has been identified that appears to function through inhibition of protein phosphatases. Secondly, the PCS phosphatase has recently been identified as a target of the transforming proteins of two DNA viruses, polyoma and SV40. The two critical links to transformation are discussed below.

Okadaic acid like the phorbol esters described in Chapter 5 is a potent tumour promoter in animal models and has pleiotropic effects on cells in culture. Of great significance is the observation that it is an inhibitor of the PCS phosphatases and to a slightly lesser extent of the AMD-phosphatases. The PCS phosphatases have a very high affinity for okadaic acid (K_i values below nanomolar) so that the PCS phosphatase can be considered as the intracellular "receptor" for okadaic acid. The AMD phosphatase is inhibited by concentrations of about 100 nM. Calcineurin is much less sensitive and the Mg^{2+}-dependent phosphatase is not inhibited at all.

As with any inhibitor, there is always a question with respect to specificity, i.e. is a particular enzyme the physiological target for a given inhibitor? In the case of okadaic acid all the evidence suggests that the phosphatases are indeed the principal targets. For example, okadaic acid causes an increase in the phosphorylation of many proteins *in vivo*, however, no kinases have been found to be influenced by this agent.

The ability of okadaic acid to elicit complex responses in cells and to function as a potent tumour promoter suggests that under normal physiological conditions the protein phosphatases suppress (or at least limit) a variety of responses, including growth. In this light, it may be anticipated that deletion of phosphatase genes may contribute to carcinogenesis. This would have a parallel in the deletion of the tumour suppressor gene retinoblastoma (see Chapter 7).

Further evidence of a role for protein phosphatases in cellular growth control has been obtained through the identification of normal cellular proteins that associate with tumour antigens. The small t antigen of the SV40 (papova) virus and the small t and middle T antigens of the polyoma virus form stable complexes with a number of cellular proteins; recently several of these proteins have been identified. Using T antigen specific antisera, two prominent polypeptides of molecular mass 36,000 and 63,000 were found to be coprecipitated from infected cells. Subsequently both peptides were purified and subjected to protein sequence analysis. This analysis revealed that the two proteins corresponded to the catalytic and regulatory subunits of the PCS phosphatase.

Middle T has previously been identified as the transforming protein of polyoma virus. Analysis of mutations of polyoma virus small t and middle T proteins has suggested that the interaction with the phosphatase is necessary to the transforming activity of the T antigen. However, it should be noted that certain transformation-defective mutants of middle T proteins exist that are still able to complex with the PCS phosphatase. Thus this interaction is necessary but not sufficient for transformation. It will be important to determine the consequence of middle T antigen association with the PCS phosphatase. It is plausible that as discussed above for okadaic acid, middle T may inhibit (or sequester) the phosphatase such that critical growth regulatory elements become hyperphosphorylated. Such a role would further the idea of a suppressor function for these phosphatases.

7

Regulators of transcription

1. Transcription is a focal point in genetic regulation

The differentiated cells of a eukaryotic organism contain the whole repertoire of genes which is already present in the zygote, but only a fraction of these genes is expressed at any stage in development, in a specific cell type or at a specific stage in the cell cycle. During the process of growth, development and differentiation cells respond to intra- and extracellular cues in a variety of ways frequently resulting in altered patterns of gene expression. Although the regulation of gene expression can occur at several different levels, it appears that initiation of transcription represents the most crucial regulatory event.

In the following sections we will present examples of situations where genes are transcriptionally regulated, the experimental approaches that have been used to identify and characterize such genes and models of how transcriptional regulators might work at the molecular level.

1.1 Genes are differentially expressed during development

The fundamental importance of transcriptional control is best exemplified by the identification of genes that are differentially expressed during development. Normal development requires the coordinate expression of a vast number of structural genes in a concerted fashion. One of the first known examples of differential gene activity during development is the changing pattern of chromosomal puffs in *Drosophila* and *Chironomus* larvae. It is now clear that changes in gene expression accompany the development of every organism. Understanding development will obviously require the identification of genes which are differentially expressed during development and the identification of gene products which mediate such regulation. Using a variety of approaches including "subtraction cloning" technique, a number of genes specifically expressed at particular stages of development, differentiation or growth (figure 1) have been identified.

The changes in pattern of gene expression during development require sets of genes to be coordinately regulated. It is therefore likely that there exist genes whose products act hierarchically to form a controlling network that ensures the proper temporal and spatial pattern of gene expression. Genetics has proved to be instrumental in identifying controlling

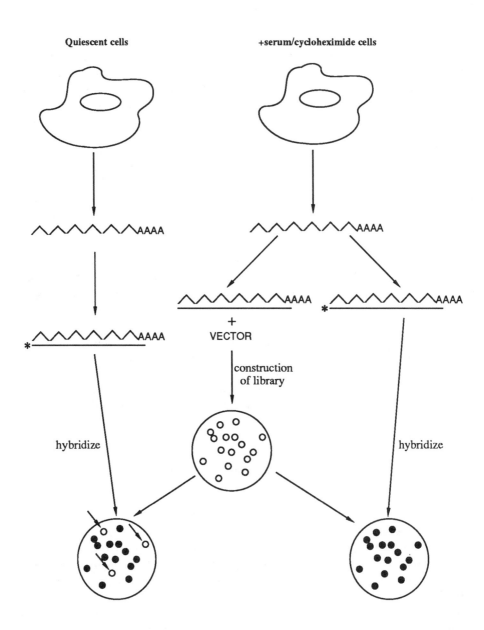

Fig 1. Differential screening of cDNA libraries identifies differentially active genes. Differential screening is used here to identify "primary response" or "immediate early" genes whose activity is induced by serum growth factors. mRNA is isolated from quiescent cells and from cells stimulated with growth factors in the presence of cycloheximide. Radioactively labelled cDNA is synthesized from both mRNA samples and used to probe a cDNA libabry constructed from the mRNA of induced cells. Clones that do not react with the probe from quiescent cells (arrows) represent genes whose transcription is induced upon serum addition. mRNA is represented by a zig-zag line with a polydenylation tail; cDNA by a solid line, and radioactive labelling is indicated by an asterisk (*).

genes that specify the body plan and the first candidates for such genes were identified early this century as homeotic mutations in *Drosophila*.

Homeotic mutations result in certain parts or an entire body segment being transformed into the corresponding structure of another body segment, thereby changing the architecture of the organism. The first homeotic genes were cloned without any previous knowledge about the biochemical properties of their gene products. The structural analysis of one such locus in *Drosophila*, the Antennapedia (*Antp*) gene led to the discovery of the "homeobox", a small DNA segment of approximately 180 base pairs, that is characteristic for homeotic genes. The significance of this DNA segment encoding a domain of about 60 amino acids, the "homeodomain", became obvious since previously unknown homeotic genes from *Drosophila* could be isolated by using the homeobox sequence as a hybridization probe. By the same approach about 50 homeobox-containing genes were isolated from vertebrates, more than 20 from human DNA, underlining the strong evolutionary conservation of this structure. Additionally, the expression of these genes show remarkably similar spatial expression patterns during development of invertebrate and vertebrate embryos. As will be discussed below the homeodomain allows proteins that contain this sequence to bind specifically to DNA and it is likely therefore that many of the homeodomain-containing proteins are transcription factors that coordinately regulate the transcription of developmentally important genes.

There is some evidence that transcriptional activation of some of these putative transcriptional activators themselves can at least be partly regulated by morphogens such as retinoic acid. If retinoic acid directly activates or represses the transcription of homeobox genes, it would presumably do this via the retinoic acid receptor, itself a transcriptional activator which binds DNA via the "zinc finger" motif as discussed below.

Clearly therefore, development is largely controlled through the action of a number of transcriptional regulators that constitute one or more regulatory cascades. A major goal of the future is to identify the final target genes of such cascades since they will undoubtably include genes important for the establishment of embryonic patterns.

1.2 Regulation of cell type-specific genes

Similar experimental strategies to those used in the study of developmental genes have been used to identify genes and gene products which specify the characteristics of a particular cell type. Differential cDNA library screening has led to the identification of genes that are differentially regulated in precursors and the differentiated representatives of a particular cell type. This approach has led to the discovery of the muscle-specific genes: Myo D, Myf-5, myd and myogenin which appear to constitute a family of genes that interact to regulate the determination and differentiation of muscle cells.

When mouse embryonic cells were treated with 5-azacytidine, an agent interfering with DNA-methylation *in vivo*, they differentiated into myoblasts. Based on the frequency of this conversion and supporting genetic evidence, it appeared that only a limited number of genes would be intimately involved in regulating this change. In order to identify such genes subtraction cloning and differential screening were used to clone genes which were exclusively expressed in myoblasts but not in the parental embryonic cells. This led to the identification and isolation of Myo D1 (Myoblast Determination gene number 1). Apart from being a muscle-specific gene Myo D1 also turned out to have a triggering function in the conversion of mouse embryonic cells into myoblasts. Upon transfection of Myo D cDNA into fibroblasts they stably convert into myoblasts. Myo D1 is a nuclear phosphoprotein, which is only expressed in proliferating myoblasts and differentiated myotubes but not in non-muscle cells. While Myo D1 represents a gene induced upon differentiation of precursor cells into muscle cells, the finding that Myo D1 carries a DNA-binding domain suggested that it may itself be responsible for inducing the muscle differentiation program by activating transcription from other promoters. Indeed, recent studies have shown that Myo D protein can participate in the transcriptional activation of other muscle specific genes. This is one of many examples where the study of cell type-specific gene expression has demonstrated that such specific gene expression can be mediated by transcription factors that are either exclusively expressed in the respective cell type or that can only function in the particular cell type.

1.3 The mitogenic response is accompanied by changes in gene expression

Another key example where transcriptional control has been shown to play a vital role in an important biological response is the growth stimulation of quiescent cells. Stimulation of such cells by growth factors or other mitogens results in the transient stimulation of transcription of a large number of genes. The ultimate fate of mitogenic signal transduction is to initiate a cascade of nuclear functions that culminate in cell division.

By differential screening protocols (see Figure 1) which compare gene expression in quiescent and stimulated cells at least 100 such genes have been identified and are collectively known as "immediate early" or "primary response" genes; they include a number of proto-oncogenes emphasizing the importance of these genes in the control of cell growth (Table 1). The genes can be grouped according to their cellular localisation and function of their products; many of the nuclear encoded products are known transcriptional regulators. Thus the stimulation of quiescent cells involves a temporal cascade of transcription stimulation: the initial response is the stimulation of transcription of the "immediate-early" genes some of which in turn stimulate transcription of other important growth-related genes. The initial stimulation does not require new protein synthesis and must therefore be meidated by pre-existing transcription factors whereas the subsequent stimulation is dependent upon the synthesis of new proteins.

2. Induction of transcription needs promoters, responsive elements and DNA-binding proteins

Although the mechanisms and biochemical pathways which cells use to translate physi-ological cues into transcriptional changes are still largely unknown, it is clear that the frequency of initiation of mRNA synthesis depends ultimately on *trans*-acting protein factors that interact with specific DNA elements in gene promoters.

Gene	Other isolates	Possible function
c-*myc*	-	transcriptional modulator
JE	N65	cytokine
TCA3	-	cytokine
c-*fos*	TIS28	transcriptional modulator
fra-1	-	transcriptional modulator
fos-B	-	transcriptional modulator
egr-1	TIS8/KROX24/ zif-268/NGF1A	transcription factor
KROX20	*erg*-2	transcription factor
NGF1B	TIS1/nur 77	ligand dependent transcription factor
c-*jun*	-	transcription factor
*jun*B	-	transcription factor
*jun*D	-	transcription factor
TIS7	PC4	cytokine
TIS11	-	not known
NS1	-	cytokine
mtf	-	not known
TT13	helix destabilizing protein	transcription initiation
AF21	C-rel	not known
N51	gro/MGSA	cytokine
A15	plasminogen activator inhibitor	wound healing
C15	tissue factor	wound healing
B2	β-actin	cytoskeletal
TTI	fibronectin	wound healing
V58	α-tropomyosin	cytoskeletal

Table 1. Some examples of primary response genes induced by growth factors. (According to Herschman/1989/ TIBS 14, 455).

In contrast to bacteria, where a promoter for efficient transcription can essentially be formed by two conserved DNA motifs located -35 and -10 bp upstream of the mRNA initiation site, in higher organisms the process is more complex. In many eukaryotic genes transcribed by RNA polymerase II (pol II) a short sequence motif consisting of A and T nucleotides, the TATA box, is located about 30 bp upstream of the mRNA initiation site. The TATA box determines the point of transcription initiation. The control region in the immediate vicinity of the transcription start site is called the promoter. Additional promoter elements, located upstream and downstream from the TATA box, are commonly found in genes transcribed by pol II. These *cis*-acting elements are usually arrayed within several hundred base pairs of the initiation site and serve as binding sites for transcription factors. Additional DNA elements can exert control over a much larger distance (1-30 kb) and in an orientation-independent fashion. Such elements or regions are called enhancers and they were first discovered in viral genomes.

The ability to modify DNA sequences and introduce DNA into mammalian cells by DNA transfection protocols, has led to the identification of the *cis*-acting control elements in a number of different promoters. If the promoters are linked to reporter coding sequences that specify readily assayable products, then the effect of promoter mutations on gene expression can be easily measured.

Having defined a promoter control region, the precise DNA sequence which constitutes the binding site for cellular *trans*-acting or transcription factors is then established by the technique of DNase footprinting (figure 2). This technique identifies DNA sequences that are protected from digestion with DNaseI as a result of interaction with a DNA binding protein.

Systematic mutational analysis has revealed that each gene in an animal cell has a particular combination of positive and negative regulatory *cis*-elements that are uniquely arranged as to number, type and spatial array. Although the interplay between the various factors that bind to these sequences is not yet understood, it is believed that combinations of *cis*-elements arranged in unique configurations confer on each gene an individualized spatial and temporal transcription program.

In summary therefore the expression of a particular gene is dependent upon an array of *cis*-acting DNA elements, consisting of enhancers and responsive elements besides the general transcription control elements. These elements are recognized by specific *trans*-acting protein factors. It is the nature of this array that determines the pattern and extent of expression.

2.1 *Trans*-acting factors contain binding domains and activation domains

Two types of *trans*-acting factors are important for efficient transcription from a given promoter (figure 3). One group, the general transcription factors, are capable of mediating transcription initiation by themselves albeit at a low level. They are required for the transcription of all genes and include RNA polymerase II and a number of factors that appear to be closely associated with the polymerase. Included in these is a factor called TFIID, which binds to the TATA-box sequence element. The second class of factors bind specifically to the upstream *cis*-acting binding sites and as discussed above, each promoter contains a specific array of such sites. The exact role of these specific factors is not clear but they mediate an enhanced rate of transcription initiation; in some cases the enhancement is constitutive whereas in other cases occurs only in some cells or in response to certain stimuli mediated through the appropriate response element.

A major advance over the last few years has been the identification of the proteins that specifically bind to the regulatory DNA elements and in many cases the genes that encode them have been cloned. Isolation of these proteins, which are usually present at low abundance in the cell, can be achieved by sequence-specific DNA affinity chromatography using a synthetic copy of the control element as target. If sufficient material can be obtained, specific antibodies to the protein can be raised or peptide sequence obtained. Such materials or sequence information can then be exploited to clone the respective cDNAs. An

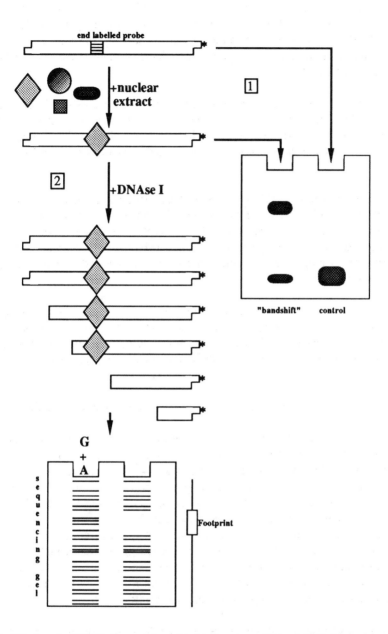

Fig 2. Identification of a DNA-binding activity and its binding site on DNA. A DNA fragment containing a transcriptional control region is radioactively labelled at its 3'-terminus and then incubated with nuclear extract containing potential binding proteins. [1] After incubation the samples are electrophoresed through non-denaturing polyacrylamide gels. DNA fragments carrying bound protein(s) show decreased mobility ("bandshift") compared to the "free" probe. [2] Incubation of the labelled probe in the presence of nuclear extract can be followed by a mild DNaseI digestion and the resulting fragments visualised by electrophoresis and autoradiography alongside the base-specific reaction products of the Maxam-Gilbert seuqencing method. Areas on the fragment which are protected from DNase digestion by bound proteins (footprint) appear as an absence of cleavage products in the DNA sequence.

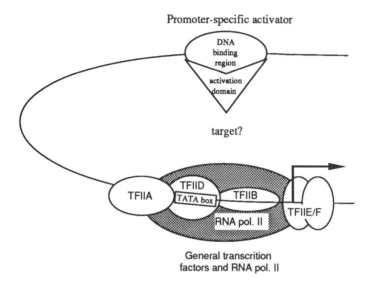

Fig 3. General and specific transcription factors. Two categories of transcription factor are illustrated. The lower group represent the general factors required for transcription. The upper factor represents the specific binding of a transcription factor to a distal site; as a consequence of this binding the activation domain interacts with an element of the general transcription apparatus to enhance initiation of transcription.

alternative strategy to identify clones encoding DNA-binding proteins has recently been employed with great success. The method involves screening of a cDNA expression library with a specific DNA probe containing the DNA-binding region of interest.

The cloning of cDNAs that encode such transcription factors has allowed detailed characterisation of the regions of these factors important for their function, in particular the identification of crucial amino acid residues involved in DNA binding and transcription activation. Some general features have emerged from such analyses and different types of protein motifs that mediate such function have been described. One surprising but common feature is that the DNA binding and activation domains are usually separable and of modular nature. This means that they can be replaced by the corresponding region of an unrelated factor and are functionally independent of their position.

Most commonly the DNA binding properties of the factors are assayed *in vitro* by the electrophoretic mobility shift assay (EMSA) (alternatively called the band-shift or gel retardation assay), where free radioactively labelled DNA is separated from DNA-protein complexes based on their different electrophoretic mobilities in polyacrylamide gels (see figure 2).

The DNA binding domains of several factors have been shown to be small autonomous domains of 60-100 amino acids. A number of distinct structural motifs can be distinguished that are used by factors to bind DNA (figure 4).

Zinc fingers

Zinc finger motifs were originally identified as DNA binding structures in the RNA polymerase III transcription factor (TFIIIA). This factor binds to the control region of the 5S RNA gene and carries nine tandem sequences of the form $Cys-X_2-Cys-X_{12}-His-X_3-His$ where X is any amino acid. The motif consists of about 30 amino acids with 2 cysteine and 2 histidine residues that stabilize the domain by tetrahedrally coordinating a Zn^{2+} ion (figure 4). This type of motif is used by a number of factors that regulate RNA polymerase II mediated transcription. For example, the mammalian transcription factor Sp1 which binds to the so-called GC-box (GGGCGG) displays three tandem TFIIIA - like zinc fingers at its C-terminus, which have been shown to be necessary and sufficient for DNA binding in the presence of zinc ions.

A second class of zinc finger motif is exemplified by the steroid hormone receptors. This motif uses 2 pairs of cysteines rather than the cysteine-histidine arrangement of Sp1 (figure 4).

It is believed that the highly conserved amino acids in the finger confer a structural framework for the binding domain. The determinants of binding specificity, however, must lie elsewhere, since the zinc finger DNA binding proteins exhibit a variety of DNA sequence specificities. For example the progesterone and the estrogen receptors show slightly different binding specificities which have been found to be due to differences in several non-conserved amino acids at the base of the finger region.

The homeodomain

A different type of DNA binding domain is the homeodomain (HD). This domain encompasses about 60 amino acids, about 30% of which are basic residues. As described above this motif was first identified as a conserved protein segment in several regulators of *Drosophila* embryogenesis and was soon found in genes of vertebrate organisms as well. The HD sequence is distantly related to the helix-turn-helix DNA binding structure of prokaryotic repressors. These prokaryotic proteins bind as dimers to dyad symmetric DNA sequences. The resolution of their crystal structure revealed that they possess two alpha-helices separated by a relatively sharp beta-turn. The more C-terminal helix from each of the two subunits (termed helix 3) is also called the recognition helix since it is directly involved in contacts with the DNA and therefore most important for sequence-specific binding. In contrast to bacterial helix-turn-helix proteins an additional helix, called helix 1, is predicted for the homeodomain. NMR analysis of purified Antennapedia homeodomain protein has directly supported the helix-turn-helix structure.

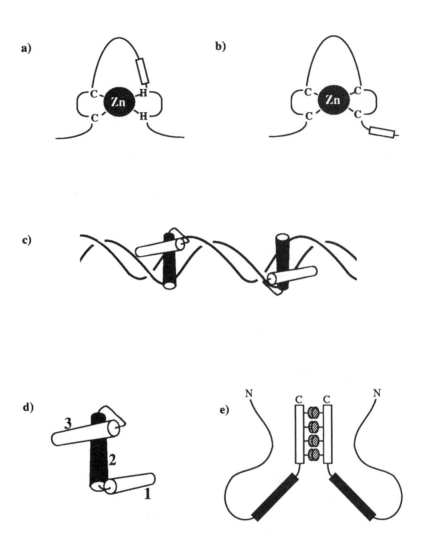

Fig 4. DNA-binding domains of transcriptional regulators. Schematic representation of the structural motifs described in the text. Shown are the two types of zinc finger motifs: in (a) the TFIIIA-type (2 cysteines-2-histidines), in (b) the steroid receptor type (4 cysteines), coordinating a zinc ion (black circle). The regions of these zinc finger proteins proposed to be invoved in specific contacts to DNA are shown as boxes. (c) A prokaryotic helix-turn-helix protein bound to a dyad-symmetric binding site on the DNA. The recognition helix (helix 3) is represented as a light barrel situated in the major groove of each half of the dyad-symmetric binding site. Helix 2 is situated above helix 3 in a position that helps lock the recognition helix into place. (d) The homeodomain is shown schematically as an extension of the helix-turn-helix motif with the additional helix (helix 1). (e) Two polypeptide chains of a leucine zipper protein dimerized via hydrophobic interactions between the two alpha helixes of the leucine zipper. Protrusions at the dimerization interface represent interdigitating leucine side chains. The black boxes represent highly basic regions involved in direct contact with DNA.

Several mammalian transcription factors including OCT-1 protein (also known as OTF-1, OBP-100, NF-A1, NFIII), the B-cell-specific factor OCT-2 (also known as OTF-2, NF-A2) and the pituitary-specific *trans*-acting factor Pit-1 (GHF-1) have been shown to bind to DNA via an HD-like domain. Thus it has become clear that the homeodomain is not only involved in developmental regulation but that this structural motif is also used for cell type-specific gene expression.

The DNA binding domains of Pit-1, OCT-1, OCT-2 and Unc-86, a developmental regulatory protein of the nematode *Caenorhabditis elegans*, constitute a subclass within the HD-containing protein family. These factors all contain a conserved bipartite domain of about 160 amino acids, which consists of an HD and a second region, the POU box. The whole domain has been termed the POU domain (Pit, OCT, Unc).

The leucine zipper

A third type of DNA binding domain was first described for the mammalian CAAT/Enhancer Binding Protein C/EBP. This domain is found in a number of transcription factors from yeast, plants and mammalian cells including the products of the oncogenes *jun*, *fos* and *myc* and a group of factors that mediate cAMP stimulation of promoters (the CREB/ATF family, see below). The domain is characteristically a stretch of about 50 amino acids with two N-terminal clusters of basic amino acids immediately followed by a region containing four or more leucine residues positioned at intervals of 7 amino acids (figure 4).

This heptad array of leucines (termed the leucine zipper) is reminiscent of the heptad repeat common to proteins that adopt a coiled-coil quaternary structure (e.g. tropomyosin, keratins, lamins) which usually results from the coiling of two parallel alpha-helices around one another. It has been suggested that polypeptides carrying the basic-leucine zipper domain (bZip) and which bind to DNA as dimers are held together by the interdigitation of two alpha-helices one from each of the monomers. Mutational analysis of a number of proteins containing this motif confirmed this hypothesis and additionally showed that the leucine zipper was crucial for dimer formation whereas the adjacent basic region made contact with DNA.

The helix-loop-helix

An amino acid comparison of several regulatory proteins, the achaete-scute and daughterless gene products of *Drosophila*, the muscle determination factors Myo D and myogenin, three *myc* encoded proteins and two immunoglobulin K chain enhancer binding proteins E12 and E47 revealed the existence of a novel DNA binding motif, the "helix-loop-helix" (HLH) motif. This motif consists of 2 amphipathic helices separated by a loop region of variable length. Adjacent to the first helix is found a basic region which is important for DNA binding. The HLH motif like the leucine zipper, is crucial for dimer formation. Thus for many factors the protein requirements for binding to DNA are two-fold: first, a basic region that contacts DNA, and secondly a region that mediates dimer formation. Mutation of either domain results in loss of DNA binding.

Other DNA binding domains

The primary sequences of other recently cloned transcription factors indicate that the number of DNA binding domain motifs is not limited to the four described above but that there may well be as yet unidentified DNA binding motifs. For example, a factor (UBF) that binds to the ribosomal RNA gene promoter shares significant homology within its DNA binding region to the non histone proteins HMG1 and 2, and to a number of yeast DNA binding proteins; the related region has been termed the HMG box. Another example is the DNA binding domain of the transcription-replication factor CTF/NF-I, which has been localized to the N-terminal third of the protein by deletion analysis. This region could form an alpha-helical structure and has a high density of basic amino acids consistent with a DNA binding region, but the protein shows no characteristic features associated with zinc fingers, homeodomains, or leucine zippers.

Activation domains

The ability of *trans*-acting factors to achieve transcriptional activation depends on regions of as few as 30 to 100 amino acids that are separate from the DNA binding domain. Activation domains are either identified by mutational analysis or by pairing a DNA segment containing a putative activation domain with heterologous DNA binding domains to produce chimeric transcription factors which are assayed *in vitro* or *in vivo* for their transcriptional activity. Factors often have more than one activation domain, and several apparently unrelated structural motifs have been identified that can confer activation. The first defined activation regions in eukaryotic transcription factors were identified in the yeast factors GAL4 and GCN4. These activation domains consist of relatively short stretches of amino acids with apparently only 2 common features: they possess a significant negative charge and can form amphipathic alpha-helices. When linked to a heterologous DNA binding domain, these domains can activate transcription of reporter genes with a binding site for the heterologous factor in yeast and in cells of a variety of higher organisms. Some mammalian factors contain such activation domains. For example, two activation domains have been identified in the glucocorticoid receptor, one of which fits the description of an alpha-helical structure. A similar correlation seems to hold true for the *c-jun* protein-activation regions.

As more and more factors are structurally and functionally dissected however, it is becoming clear that the "acidic" activation domain is not the only type of structure that can mediate transcription activation. For example, the major activation domains of the Sp1 factor are rich in glutamine residues whereas the activation domain of the CTF factor is proline-rich (figure 5).

At present it is unclear how these various domains activate transcription. It has been proposed that the three different types of activation domains identified so far (acidic, glutamine-rich, proline-rich) are likely to function by contacting other proteins, either general transcription factors associated with pol II like the TATA box binding protein TFIID or different subunits of pol II itself. Since specific association between the regulatory regions of active genes and nuclear scaffold proteins has been demonstrated, some activation domains might contact proteins of the nuclear matrix, thereby tethering

GAL4 acidic domain

...D SAAAHHDNST I PLDFMPRDALHGFDWSEE DDMSDGLPFLKTDPNHNGF...

Sp1 glutamine-rich domain B

...QGQTPQRUSGLQGSDALNI QQNQTSGGSLQAGQQKEGEQNQQTQQQQI
LI QPQLUQGGQALQALQAAPLSGQTFTTQAISQETLQNLQLQAUPNSG
PI I IRTPTUGPNCQUSWQTLQLQNLQUQUPQAQTITLAPMQGUSLGQ...

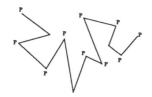

CTF proline-rich domain

...P PHLNPQDPLKDLUSLACDPASQQPGALNGSGQLKNPSHCLS
AQNLAPPGLPALALPPATKPATTSEGGATSPSYSPPDTSP...

Fig 5. Activation domains of transcriptional activators. Three examples of protein domains which confer transcriptional activation of DNA-binding factors are shown. The amino acid residues that are characteristic for the three domain types are shown in bold: aspartic acid and glutamic acid (D,E) for the acidic domain, glutamine (Q) for the glutamine-rich, and proline (P) for the proline-rich domains. The examples are: acidc domain, yeast facotr GAL4; glutamine-rich, transcription factor Sp1; proline-rich, CTF/NF1. (According to Mitchell and Tjian/ 1989/Science 245 371).

genes to nuclear regions with locally high concentrations of other essential transcription and processing factors. Binding sites of selected *trans*-acting factors are summarized in Table 2.

Factor		Binding site (5'-3')
GR	(glucocorticoid dependent transcription factor)	GGTACAN3TGTTCT
Sp1	(three zinc finger DNA binding protein)	GGGCGG
CTF/NF-I	(proline-rich transcriptional activation domain)	GCCAAT
c-jun	(member of AP-1 site binding family)	TGTGGAAAG
C/EBP	(leucine zipper associated DNA binding protein)	TGTGGAAAG
AP-2	(AP-2 site binding protein)	CCCCAGGC
CREB	(confers cAMP inducibility and E1A transactivation)	TGACGTCA
OCT-1	(ubiquitous Pou-domain DNA binding protein)	ATTTGCAT
OCT-2	(Pou-domain and leucine zipper DNA binding protein)	ATTTGCAT
SRF	(serum response element binding protein)	GATGTCCATA-TTAGGACATC

Table 2. Binding sites for some DNA-binding proteins.

2.2 Some responsive elements bind families of transcription factors

The TPA-responsive element (TRE) was identified as a *cis*-regulatory element that allowed promoters to be stimulated by phorbol esters. The factor that recognized this element was identified as AP-1 (transcription activator protein 1), a factor already implicated in the activation of viral enhancers. When the structure of AP-1 was resolved it turned out to be very complex, consisting of more than one polypeptide. One of these was found to be closely related or identical to the product of the oncogene *jun*, a *trans*-acting factor containing a leucine zipper dimerization domain. It was subsequently found however that *c-jun* protein bound the TRE inefficiently as a homodimer. In contrast, when dimerized with the product of the *fos* oncogene, it bound very efficiently. The difference in binding efficiency reflects a large difference in the stability of the dimers, the *jun-fos* protein heterodimer being 500-1000 fold more stable than the *jun-jun* protein homodimer. The *jun* and *fos* genes both belong to families of genes several members of which (*junB*, *junD*, *fra1*, *fosB*) have been characterized and cloned (figure 6). Any member of the *jun* family can heterodimerize with any member of the *fos* family and the different combination of dimers appear to bind the TRE with equal specificity and efficiency *in vitro*. The reason for such complexity is presently unclear but it is conceivable that the different dimers can respond to different internal or external signals.

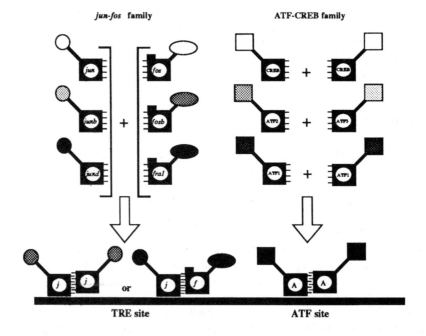

Fig 6. Dimerization within transcription factor families. Members of the *jun* family can weakly homodimerize or form stronger heterodimeric complexes with members of the *fos* family. *fos* family members do not homodimerize. The *jun/fos* heterodimers bind equally well to the TPA responsive element (TRE) *in vitro* but different complexes may have different transcription regulatory properties. The ATF-CREB family appear to obey more complex dimerization rules. Specific homo-and heterodimers form and they all bind to the ATF site. (According to Ziff/ 1990/TIGS 6 69).

Dimerization between *fos* and *jun* occurs via the leucine zipper domains of both proteins. Zipping juxtaposes the basic regions of *fos* and *jun* and accounts for cooperative binding of the *fos-jun* heterodimer to the dyad symmetric DNA sequence found in the palindromic TRE. This model also explains the observation that molecules that are unable to dimerize cannot bind DNA, since the DNA-binding domain requires structural contributions from both subunits of the dimer. The zippers are also responsible for a high specificity of dimerization: the *fos* protein efficiently dimerises with the *jun* protein zipper but not with itself (or many other proteins that contain zippers). The rules that govern specificity of dimerization are presently not clear.

The cAMP-responsive element (CRE) was identified in the rat somatostatin gene promoter as a highly conserved palindrome which confers inducibility by cAMP. The consensus sequence of the CRE (5´-TGACGTCA-3´) differs from the TRE by the insertion of a central C residue; this sequence is recognized by the CRE binding protein CREB, also known as ATF (Activator Transcription Factor).

At least eight cDNAs have been cloned that encode CREB/ATF binding proteins. They all contain a bZip DNA binding motif; beyond this they are completely different. The CRE/ATF binding site has been implicated in mediating activation of transcription not only by cAMP but also by two viral encoded proteins, namely the E1A protein of adenovirus and the *tat* protein of the human T-cell leukemia virus HTLV-1. It is possible that different members of the CREB/ATF family mediate these different responses. It is very likely that many of the DNA regulatory or responsive elements bind families of factors rather than a single protein. Although the reasons for this are not clear, it obviously provides for a great deal of scope and flexibility in the regulatory process.

3. Regulation of *trans*-acting factor activity

The previous chapters of this book have described transmission of an extracellular signal across the plasma membrane and through the cytoplasm. In this chapter so far we have presented evidence that changes in gene expression are critical for the cellular response to many of those extracellular and intracellular signals and that these changes in gene expression require the cooperation of *trans*- and *cis*-acting control elements. How do these transduced signals activate transcription? The induction of some genes does not require protein synthesis and therefore the mediators of the signal have to be already present in the cell. The activity of these mediators or factors must therefore be altered in response to the transduced signal. Proposed mechansims to account for factor activation include the binding of ligands to the factor, post-translational modification of the factor such as phosphorylation and modulation of the association of the factor with other proteins. Although these mechanisms do not necessarily operate independently of each other, examples of each type will be described separately. Regulation of genes that are expressed at later stages can be achieved by *trans*-acting factors that have been synthesized, i.e. induced earlier during the cellular response. If such a factor is limiting increased synthesis will lead to increased transcription from promoters with which it specifically interacts.

3.1 Some regulators of transcription are intracellular hormone receptors

One group of receptor molecules, including those for steroid hormones, vitamin D3, juvenile hormone, thyroid hormones and retinoic acid is not localized at the plasma membrane but inside the cell. The ligands for these receptors are membrane permeant. These receptors are DNA binding proteins and the ability of the receptors to stimulate transcription by interacting with specific DNA sequences is dependent on ligand binding. Therefore, these receptors can be described as ligand-regulated *trans*-acting factors.

Ligand binding to the receptor appears to have several possible consequences including the promotion of receptor dimerisation, binding to DNA and localisation of the receptor to the nucleus. In addition some transactivation domains of the receptor appear to require the ligand. All known members of the nuclear receptor superfamily are roughly similar in their modular organisation (figure 7). They exhibit a variable N-terminal region, a short and well conserved cysteine-rich central domain necessary for DNA binding and a relatively well conserved C-terminal hormone-binding region.

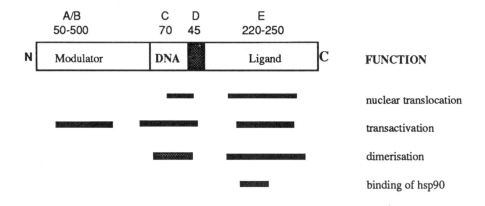

Fig 7. The super family of nuclear receptors. Alignment of the amino-acid sequences of different nuclear receptors (steriods, retinoids, thyroid hormone) shows that they are composed of a series of conserved domains designated as A/B, C, D and E. Involvement of each of these domains in specific functions of the nuclear receptors is indicated by solid bars.

The DNA binding domain of the steroid receptor family possess a cysteine-rich domain of the form Cys-X_2-Cys-X_{13}-Cys-X_2-Cys-X_{15-17}-Cys-X_5-Cys-X_9-Cys-X_2-Cys-X_4-Cys which can form two putative zinc fingers and which determines DNA binding specificity and requires zinc or cadmium ions for binding. Each of the hypothetical fingers is encoded by a separate exon and both fingers are required for binding to the hormone response element (HRE). The sequences of the HREs are palindromic and they are recognized by receptor dimers. Site-directed mutagenesis and experiments in which domains were swapped between different receptors revealed that the two zinc fingers of the receptor monomer have different functions in DNA binding. The N-terminal finger determines target gene specificity via amino acids at the root of the finger whereas the C-terminal finger contacts the sugar-phosphate backbone of the HRE-flanking sequences. The 10 amino acids following the first zinc finger have the potential to form a strongly amphipathic α-helix, therefore this region may form some type of recognition helix. Interestingly, a single amino acid substitution in this region abolishes transcriptional activation without interfering with DNA-binding activity and specificity, suggesting that DNA-binding is not sufficient for transcriptional activation. A genetic defect leading to vitamin D resistant rickets in two different families also shows that each of the two hypothetical zinc fingers is physiologically important. A single point mutation in the vitamin D3 receptor gene in both cases leading to the exchange of a conserved amino acid in the first or second finger inactivates the vitamin D3 receptor and makes the patients vitamin D resistant.

3.2 Transcription factor activity may be regulated by phosphorylation

Since many transcription factors are phosphoproteins, their activity might be regulated by phosphorylation-dephosphorylation events. A prominent example for such regulation is the heat shock transcription factor. The genes encoding heat shock proteins and the mechanisms of their transcriptional induction by elevated temperature or other types of stress are highly conserved in eukaryotes. The induction is mediated by a *cis*-regulatory sequence known as the heat shock element (HSE). This element is recognized by a DNA binding transcription factor, the heat shock transcription factor HSF or HSTF. Although HSFs from different species (yeast, *Drosophila*, human) all recognize the same HSE, the manner in which their activity is regulated appears to be different from yeast to higher eukaryotes. In yeast the HSF is present in unshocked and shocked cells in a form that readily binds to DNA. However, heat shock increases the phosphorylation state of the factor which correlates with increased transcription from the HSE. In *Drosophila* and human cells, however, HSF DNA binding activity is undetectable at normal temperature. Upon heat shock the affinity of HSF for its binding site is drastically increased. It is likely, although not yet proven, that phosphorylation is involved in this unmasking of binding activity. Phosphorylation also appears to play a role in regulation of the activity of the cAMP response element binding factor CREB. Specifically, phosphorylation at a site recognised by the cAMP dependent protein kinase is required for its ability to activate transcription.

3.3 Protein-protein interactions can also control transcription factor activity

Another mechanism that is employed to regulate the activity of certain transcription factors involves protein-protein interaction. This type of regulation is exemplified in the activation of the glucocorticoid receptor upon hormone binding and the activation of the immunoglobulin K light chain enhancer-binding protein NF-KB following phorbol ester treatment. In both cases in the absence of inducing signals the inactive transcription factor is found in the cytoplasm in association with another protein. Protein synthesis is not required for activation in either case. Inactive glucocorticoid receptors are associated in a large multi-protein complex that includes the heat shock protein, hsp 90, which is abundant in the cytoplasm. As a result of hormone binding to the receptor the complex dissociates and nuclear localization signals within the receptor become unmasked and direct the receptor to the nucleus where it binds to the appropriate response elements and activates or represses transcription of adjacent genes. The complex formation between glucocorticoid receptor and hsp 90 is mediated through the hormone binding domain and the hormone binding domain is sufficient to confer steroid-dependent regulation on other proteins.

The transcription factor NF-KB was originally identified as a B cell-specific DNA binding protein that recognizes a *cis*-regulatory element within the enhancer of an immunoglobulin light chain gene. NF-KB is sequestered in the cytoplasm by association with a cytoplasmic inhibitor IKB. Phorbol ester treatment of B cells leads to the appearance of NF-KB DNA binding activity which can be mimicked in vitro by phosphorylation of the inhibitory protein IKB with protein kinase C, the mediator of phorbol ester action. NF-KB, initially believed to be exclusively expressed as a lymphoid-specific factor has since been found in the cytoplasm of non-lymphoid cells where its DNA binding activity can be unmasked by treatment with phorbol ester without concomitant transcriptional activation of lymphoid genes that contain the NF-KB binding site. This suggests that the DNA binding activity of this factor can be induced in different cell types by the same mechanism but that additional cell-specific components are then required in order to form active transcription initiation complexes.

4. Nuclear oncogenes and anti-oncogenes

On the basis of their ability to act as dominant oncogenes during transformation the products of nuclear oncogenes seem likely to play a major role in cell growth and development. Many are phosphoproteins and could therefore be the final recipients of transduced signals. A number of such transforming proteins have been found to be either transcription factors or *trans*-acting regulators of transcription that themselves do not bind to DNA. These observations clearly emphasize the key role that transcription regulation plays in the control of cell growth and division. Examples of nuclear oncogenes are discussed below.

4.1 DNA tumour viruses

The study of tumour viruses has proven to be instrumental not only in studying the process of oncogenesis but also transcription regulation. The tumour viruses can be divided into two classes depending on whether they contain a DNA or RNA genome. One of the well studied DNA tumour viruses are the Adenoviruses which are human viruses that can induce tumours in newborn rodents or morphologically transform a variety or primary cells in culture. Adenovirus contains a very powerful transcriptional regulator, the E1A gene product, that can activate or repress a variety or viral and cellular promoters and contributes to the tranforming activity of the virus.

Based on sequence similarities of E1A proteins of different adenovirus serotypes a pattern of highly conserved regions interspersed with sequences of low homology emerged. The 3 conserved sequences, each between 18 and 48 amino acids long, are designated as regions 1-3 and have been shown by mutational analysis to be functional domains of the E1A protein (figure 8).

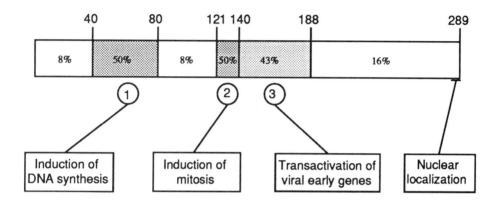

Fig 8. Structural and functional properties of the adenovirus transactivator E1A. Shown is a schematic representation of the regions responsible for E1A activity. The three conserved regions of the E1A protein are labelled 1, 2 and 3 and their functional assignments are shown.

The role of E1A as a transcriptional activator is crucial to viral growth since it is required to stimulate expression of a number of other adenovirus genes. Despite a great deal of attention, the mechanism by which E1A protein activates transcription remains unclear. It does not directly or specifically bind to DNA but rather appears to modify the activity of upstream sequence-specific binding factors. How does it do this? There are two main models: 1) E1A protein directly or indirectly stimulates the post-translational modification of certain factors, 2) E1A protein is attracted to a promoter by protein-protein interaction with certain binding factors and subsequently interacts directly with the transcription machinery. The latter model has mainly stemmed from the observation that a small domain of the E1A protein (the conserved region 3) can act as a very efficient transcriptional activation domain.

Thus although the E1A protein cannot bind to DNA, it could be brought to the promoter by interaction with another factor that provides DNA binding specificity. This model is reminiscent of the mechanism by which another viral protein has been demonstrated to activate transcription. The VP16 protein of the Herpes Simplex Virus activates transcription of a set of viral genes very early in the infection cycle. The protein contains a domain that behaves as a very strong activation domain and an additional region that allows it to specifically interact with the OCT-1 DNA binding protein mentioned earlier. This interaction therefore brings VP16 protein to certain promoters that contain OCT-1 binding sites.

The discovery that E1A protein as a transactivator of transcription led to the attractive hypothesis that its oncogenic activity resulted from its ability to activate the expression of cellular genes involved in growth control. By mutational analysis, however, this hypothesis could not be confirmed since regions 1 and 2 whose concerted action is required for transformation are both dispensable for transcriptional activation. A clue to the role of the E1A protein in transformation was provided by the discovery that a 105,000 polypeptide which coimmunoprecipitates with the E1A protien is the product of the retinoblastoma (RB) gene. In fact, E1A protein has been shown to interact with a number of cellular polypeptides, although in most cases their identity is unknown.

The significance of the interaction between E1A protein and the RB gene product ($p105^{rb}$) is supported by mutational studies which show a correlation between binding of $p105^{rb}$, which requires regions 1 and 2, and transformation. SV40 large T antigen also binds to $p105^{rb}$ and the region of SV40 large T required for the binding encompasses a region which had been identified as a region with similarity to E1A region 2. Additionally the large T antigens of polyomavirus, BK virus, JC virus and the HPV16 virus E7 protein have been shown to bind to $p105^{rb}$. The conclusion from these results is that diverse DNA tumor virus nuclear oncoproteins can induce transformation by at least one common mechanism: binding to $p105^{rb}$.

The RB gene is the best example of what is referred to as "tumour suppressor", "recessive oncogene" or "anti-oncogene", meaning that tumour formation or transformation occurs when these genes are inactivated. Since genetic studies indicate that the loss of the RB gene leads to transformation, it suggests that binding of viral nuclear oncoproteins to $p105^{rb}$ inactivates the RB gene product. It is not yet known how $p105^{rb}$ might control cell proliferation although its nuclear localisation has led to the proposition that it may regulate transcription of specific cellular genes involved in growth control.

Another candidate for an anti-oncogene is the cellular protein p53. Like RB, deletion or mutation of the p53 gene has been shown to be associated with transformation. p53 is a nuclear protein that forms a stable complex with SV40 large T and the adenovirus E1B protein (figure 9). Whereas adenovirus possesses 2 dominant oncogenes E1A and E1B which are both necessary for transformation of primary cells and may function at least in part by interacting and inactivating $p105^{rb}$ and p53, the SV40 large T antigen combines the 2 tasks of the E1 proteins in one polypeptide.

An evolutionary rationale has been proposed to explain why viral transforming proteins might need to bind the products of cellular anti-oncogenes. Adenovirus normally infects differentiated cells which are growth arrested. These quiescent cells must be stimulated into S phase for efficient viral DNA replication. The induction of cellular DNA synthesis is mediated by the E1A protein and requires regions 1 and 2, the $p105^{rb}$ binding site. Regions 1 and 2 are dispensable for efficient growth in actively dividing cells. Thus DNA viruses may have evolved proteins such as E1A and SV40 large T to inactivate the products of anti-oncogenes, thus enabling efficient viral replication in their normal host cells.

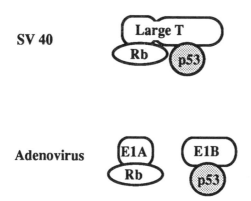

Fig 9. Complexes between the transforming proteins of the DNA tumour viruses and cellular anti-oncogene products. The large T antigen of SV40 binds both the retinoblastoma gene product Rb and the cellular nuclear protein p53. In adenovirus, the cellular anti-oncogenes are bound by two separate viral proteins: E1A binds Rb and E1B binds p53. (According to Lane and Benchimol/1990/Genes and Development 4 1).

4.2 The oncogenes *jun, fos, myb* and *erbA*

Protooncogenes, the progenitors of retrovial and cellular oncogenes, have been implicated in the control of normal growth and differentiation. Many of the products of protooncogenes and oncogenes are known to be DNA binding proteins and regulators of cellular transcription. Such genes presumably exert their effect by modulation of expression of genes intimately involved in growth control. Examples include the protooncogenes *c-jun, c-fos, c-myb* and *c-erbA*. Two crucial questions are posed by such findings: (i) what are the critical target genes whose expression is regulated by these proteins? (ii) what are the crucial differences that distinguish oncogenes from their normal protooncogene counterpart?

Studies in many different research areas are likely to contribute answers to these questions. Genes that are important players in cell cycle and cell growth control are constantly being identified and characterized. Part of the characterization involves studying their expression pattern and it is reasonable to hypothesize that some at least will be regulated by known protooncogene products. A comparison of protooncogene and oncogene products should ultimately reveal crucial differences. Such differences could prove to be subtle and for example, involve changes in efficiency or specificity. Alternatively the activity of protooncogenes may be tightly regulated during cell growth and division and that their 'activation' into oncogenes may involve an alteration in their regulation.

The products of the protooncogenes *c-fos* and c-*jun* have been identified as proteins that give rise to a DNA binding activity that specifically interacts with an element known as the AP-1 site or TRE. As described earlier, the c-*jun* protein can bind to such a site as a homodimer, albeit weakly. However, a c-*jun*/c-*fos* protein heterodimer binds very efficiently to such a site resulting in transcriptional activation. The *jun* oncogene was initially

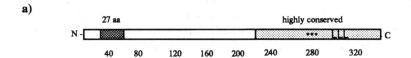

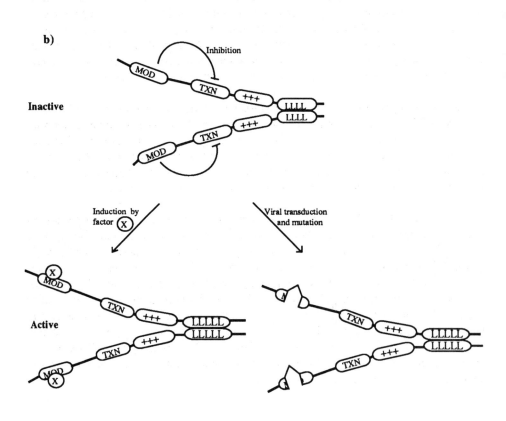

Fig 10. Model for activation of *c-jun* and *v-jun*. (a) Schematic diagram of the *c-jun* polypeptide. The C-terminus is highly conserved, containing the leucine zipper and a basic DNA-binding domain. The 27 amino acids which are deleted in *v-jun* are indicated (27aa). (b) The transcriptional activity of *c-jun* can be induced by post translational modification or by viral transduction. In the inactive state, an inhibitory modulator region (MOD) is postulated to act negatively on the transcriptional activation domain(s) (TXN) of *c-jun*. Post-translational modification of the modulator, indicated by "X", can relieve the inhibitory effect. Deletion of the inhibitory region as has occurred during retroviral transduction also relieves the inhibition. (According to Bohmann and Tjian/1989/ Cell 59 709).

identified as the transforming gene of the retrovirus ASV17, isolated from a spontaneous chicken sarcoma. Compared to its normal cellular counterpart, this viral *jun* protein has a 27 amino acid deletion that is normally important in *c-jun* protein 'activation' although the consequences of such a loss of function are not fully understood. One possible explanation has arisen from comparing the ability of *c-jun* and *v-jun* proteins to activate transcription *in vitro*; these studies suggest that the N-terminal deletion may remove a regulatory domain that negatively controls *c-jun* transcriptional activity (figure10).

The *myb* oncogene was originally discovered as the transforming gene in an avian retrovirus strain. It has been implicated in the transformation of lymphoid tumours in mice and has also been found to be amplified in a particular human tumour cell line. The *c-myb* gene contains 7 exons of which about $6^1/_2$ are transduced as the transforming gene of the avian myeloblastosis virus (AMV) and about $5^1/_2$ as the *v-myb* gene in the E26 virus. Both of these forms of the *v-myb* encoded proteins have C-terminal deletions compared to *c-myb*. These *v-myb* proteins, like the *c-myb* protein, have been shown to bind to DNA *in vitro*.

The *v-erbA* oncogene of avian erythroblastosis virus (AEV) transforms avian fibroblasts and erythroblasts. It is derived from *c-erbA* gene which encodes the thyroid hormone receptor, a member of the nuclear receptor superfamily. The thyroid hormones thyroxine (T4) and 3,5,3´-triiodothyronine (T3) which bind to this receptor are critical for the development of the central nervous system, the maintenance of homeostasis and can also influence the synthesis and activity of many important regulatory proteins.

In the genome of the ES4 strain of AEV the *v-erbA* oncogene is found alongside the *v-erbB* oncogene, a truncated version of the EGF receptor. The *v-erbA* product inhibits terminal differentiation of avian erythroblasts presumably by affecting the transcription of specific genes. The protein is a nuclear phosphoprotein truncated at the N-terminus which retains the DNA-binding but not the hormone-binding properties of the T3 receptor. It is not clear how the N-terminal deletion affects hormone binding since no function has been assigned yet to the N-terminal region of this receptor. The putative functional domains of the thyroid hormone receptors have been inferred based on structural homology with the steroid receptors. As in the steroid receptors, (see figure 7) DNA binding is mediated via the cysteine-rich region while the C-terminal region contains hormone binding function and also appears to play some role in *trans*-activation.

8

Integrated signalling

1. *In vivo* versus *in vitro*: the essential differences

The preceding chapters have described the individual components of a variety of systems employed by cells to regulate their growth and differentiation. Clearly these mechanisms are extremely complex and progress in their elucidation has been made in a reductionist fashion; that is, effort has concentrated on the investigation of pieces of pathways in isolation, rather than the whole. In attempting to rebuild a physiological system from these molecular snippets, several fundamental differences between the *in vitro* paradigm and the whole animal must be considered (but are often overlooked).

Cells in the body are not exposed to growth factors and hormones in the way in which cells in culture are in the laboratory. The environment experienced by a cell *in vivo* depends upon its physiological situation and the neighbouring tissue: thus epithelial cells are exposed to a different spectrum of conditions than mesenchymal cells or to cells of the central nervous system. Instead of being subjected to a single stimulus, cells in the body are bathed in a continual flux of signals. However, most cells in a fully developed organism are quiescent, that is, they are not continually dividing. The few cells that do multiply are tightly controlled, and often produce cells that do not have the capacity of self renewal such as erythrocytes and platelets. Other cells may be temporarily awoken from their normally refractile state in response to injury, facilitating repair and repopulation of the damage site. The reason for such tight control of the proliferative state of cells is, of course, that uncoupled growth inexorably leads to neoplasia (see section below).

The situation in a culture dish is rather different. *In vitro* cultures usually comprise dividing cells. In continuous culture, those that cannot multiply are selected against and become overgrown by cells that can. Indeed, much cell regulation research relies on the use of such "immortalised" cell lines, which have infinite lifespans in culture (see below) and may well have in part deregulated the very pathways the investigator is studying.

Responsive cells in an organism do not encounter extracellular signals in isolation: a single cell may respond to several factors simultaneously. Critically, the fate of such a cell depends on the assimilation of all of these inputs, and the combinatorial result is often very different from that obtained with each individual stimulus in a tissue culture flask.

This chapter is concerned with the ways that cells put all of the information that bombards them together. The methods by which coordination and control are achieved will be discussed, especially with respect to the most widely used mechanism of post translational control, that of protein phosphorylation. How do pathways interact? How are messages integrated in such a way that signals do not "clash"? These questions cannot yet be answered in full, but many mechanisms have been discovered, in part through the study of "normal" cells, but also from analysis of how these cellular controls are subverted in neoplasia.

2. Feed-forward and feed-back control: getting a lot out of a little and knowing when enough is enough

2.1 Feed-forward control

Although cells have tens of thousands of specific receptors, transmission of a signal may occur when only a small proportion of these receptors have bound their ligand, say with an occupancy of 10% or less; in other words, perhaps one thousand or fewer receptor molecules may be activated. In some cases, for example olfactory stimulation, the required receptor occupancy levels can be much lower. Yet these signals need to initiate a plethora of changes within the cell and do so very quickly. This signal amplification is achieved in a number of ways, all involving catalysis, where one transducing molecule catalytically stimulates the synthesis of secondary molecules, etc. Thus G-protein-mediated activation of phospholipase C generates many molecules of diacylglycerol and inositol polyphosphates, the former of which stimulate phosphorylation of many proteins through the activation of protein kinase C (see Chapter 5).

The classic example of amplification is that of glycogenolysis which acts as a paradigm for intracellular signal transduction (figure 1). In the classic fight or flight response, adrenaline binding to the β-adrenergic receptor stimulates coupling of a G-protein (G_s, see Chapter 4) to adenylate cyclase. This generates the soluble second messenger cyclic AMP which diffuses through the cytoplasm to bind to the regulatory subunits of the cyclic AMP-dependent protein kinase causing dissociation and activation of the catalytic subunit. One of the substrates for this protein kinase is another protein kinase, phosphorylase kinase. Phosphorylation of this protein stimulates its activity leading to phosphorylation of glycogen phosphorylase, which, in turn, promotes the breakdown of glycogen to glucose-1-phosphate. In this sequence of events there are five signal amplifications mediated by: (1) the β-adrenergic receptor, (2) adenylyl cyclase, (3) cyclic AMP-dependent protein kinase, (4) phosphorylase kinase and (5) glycogen phosphorylase. Such a massive increase in signal strength is required in this case to rapidly (in seconds) generate enough glucose in the muscle tissue to fuel sustained anaerobic glycolysis. The intracellular concentrations of each of the amplifiers increases as one descends the cascade; the intracellular concentration of glycogen phosphorylase being 70 μM (7 mg/ml!) while that of adrenaline is about three orders of magnitude lower.

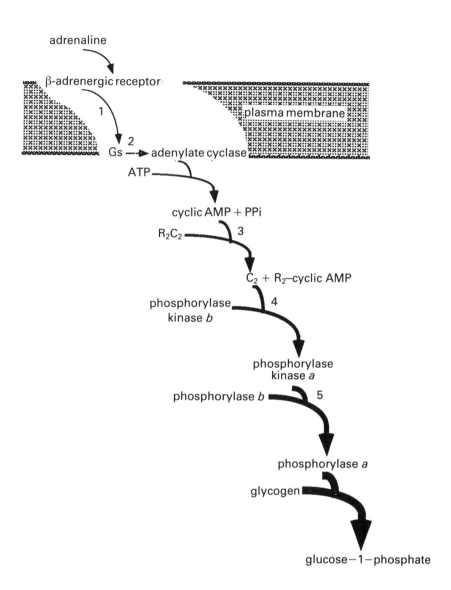

Fig 1. Adrenaline to glycogenolysis; an amplification cascade. The cascade of events leading from adrenaline binding to its receptor to increased glycogen breakdown is illustrated. This includes a series of catalytic steps that are responsible for amplification of the initial signal (see text).

The activation of one protein kinase by another also occurs in other systems including the pathway leading to ribosomal S6 phosphorylation which involves an S6 kinase, an S6 kinase kinase and probably an S6 kinase kinase kinase. Protein-serine kinases are also activated in response to a number of polypeptide mitogens acting through receptors that are protein-tyrosine kinases, such as the EGF and PDGF receptors (see Chapter 3). Such interactions between the mediators of signal transduction potentiates the incoming stimuli and forms the basis for effective coordination (see below).

Feed-forward control is also manifested in transcriptional regulation. The proto-oncogene product of c-*jun* is a component of the AP-1 transcription factor complex and can bind as a dimer to specific DNA sequences in the promoter regions of certain genes thus causing *trans*-activation (see Chapter 7). The c-*jun* protein also binds to an element in its own promoter, stimulating its own transcription (figure 2). Unlike most other immediate-early genes which are virtually undetectable in dormant cells, this autologous *trans*-activation maintains discrete levels of c-*jun* protein under all conditions. In resting cells c-*jun* protein is kept from activating transcription of other genes by being phosphorylated at sites that inhibit DNA binding; the phosphorylated protein, however, is still able to *trans*-activate its own promoter. Activation of c-*jun* protein upon hormonal stimulation is facilitated by loss of the inhibitory phosphate. This activation requires no nascent protein synthesis and is thus extremely rapid, a hallmark of early acting signal transducing systems. Subsequent induction of c-*fos* (and related genes, see below) leads to the secondary formation of heterodimers with c-*jun* protein which form more stable complexes with DNA than c-*jun protein* homodimers and are thus more potent in gene *trans*-activation. The situation is further complicated by the existence of at least four other components which competitively interact with c-*fos* and c-*jun* proteins. Many of these proteins, such as *jun*B, *fos*B, *fra*-1 and *fra*-2 are growth factor-inducible and are co-induced with c-*fos*, although with different kinetics. They can act both cooperatively and antagonistically with respect to transcriptional activation and form a temporally resolved array of complexes with discrete *trans*-activatory properties. However, the further intricacies of gene regulation afforded by this spectrum of interactions are left to the reader to contemplate.

Another form of feed forward control is that of autocatalytic activation. In this case, an initiating stimulus generates a molecule which can activate itself, thus becoming self-sufficient and impervious to the original signal. This mechanism can act as a switch; once triggered, it is self-perpetuating. An example occurs in a protein kinase found at high levels in the brain, namely the calmodulin-dependent protein kinase-II. This enzyme consists of 12 subunits and is initially entirely dependent on the presence of calcium/calmodulin for activity. However, once activated, this kinase phosphorylates itself on a threonine residue. This action relieves the kinase of its dependency on calcium/calmodulin enabling constitutive activation in the absence of effectors. What might be the function of such chronic stimulation? The duration of activation once initiated may be the lifetime of the protein kinase if no dephosphorylation takes place. Since the kinase is a multi-subunit protein, as one subunit is degraded it may be replaced by a nascent polypeptide. This would then be phosphorylated on the threonine residue thus regenerating the *status quo* and essentially immortalizing the active form of the protein. It has been postulated that such a self-renewing kinase might represent the molecular basis for memory, in which excitation of individual neurons is trapped for the lifespan of the organism.

2.2 Feed-back control

How are signals switched off? That stimuli do not have a persistent effect is demonstrated

A.   Resting cells

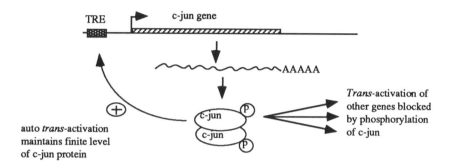

B.   Hormonal activation

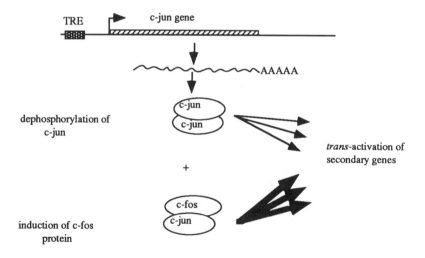

Fig 2. Regulation of c-*jun* expression and function. In resting cells c-*jun* is expressed but is restricted in *trans*-activation of other genes through phosphorylation. Following hormonal activation c-*jun* is dephosphorylated and can now transactivate other genes either alone or as a c-*jun*/c-*fos* complex.

by the phenomenon of desensitization. Upon addition of an agonist, cells respond accordingly. However, after removal of the stimulus and subsequent reapplication, a lesser response is usually observed. This ligand-induced attenuation is termed homologous desensitization since the agonist dampens down its own (homologous) signal. Conceptually, there are many ways in which this might occur. The receptor could reduce its affinity for the ligand; the ligand-induced change in the receptor that transmits the signal could be altered; the receptor could be made inaccessible to the ligand; a process downstream of the receptor could be inhibited. In fact all of these controls are used to some degree.

Exposure of growth factor receptors, including those for EGF and PDGF, to their cognate ligands causes a decrease in the number of receptors at the cell surface. This is due to down-regulation of the ligand-bound receptors by sequestration into intracellular vesicles. Internalization *per se* is not required for signal transduction and thus appears to act purely to desensitize the cell. The mechanisms involved in routing the receptors down this pathway are poorly understood. Regions of receptors that are required for intracellular sequestration have been identified by site-directed mutagenesis. In the case of the growth factor receptors, the signal for internalization is thought to be ligand-dependent dimerisation of the receptors.

The process of homologous desensitization has been most extensively studied for G-protein-linked receptors such as the β_2-adrenergic receptor system. On binding its ligand, the receptor changes conformation transmitting a signal via the activatory G-protein G_s. However, the conformational change also activates a second process since the agonist-bound receptor is a substrate for a specific protein kinase termed β-adrenergic receptor kinase (abbreviated to βARK) (figure 3). The normally cytoplasmic βARK translocates to the plasma membrane and phosphorylates an intracellular domain of the agonist-bound receptor, reducing its signalling capacity and thus attenuating the signal. Since only the activated receptor is a target for βARK, the desensitization is coincident with activation and the signal is inherently transient. In addition to the β_2-adrenergic receptor, other receptors that regulate adenylate cyclase such as the α_2 adrenergic receptor and the muscarinic cholinergic receptor are also substrates for βARK *in vitro*. There is also accumulating evidence of a role for phosphorylation in regulating agonist-dependent desensitization of other receptors such as the cyclic AMP receptor in *Dictyostelium discoidium* and the yeast α-mating factor receptor.

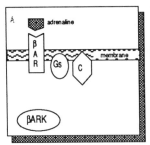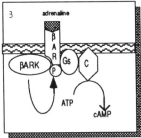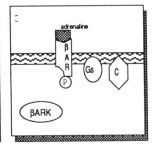

Fig 3. Homologous desensitization of the β-adrenergic receptor. A, Binding of adrenaline to the β-adrenergic receptor (βAR) causes a conformational change and both activation of Gs and translocation of the β-adrenergic receptor kinase (β-ARK). B, Phosphorylation of the receptor by the kinase uncouples the receptor from the G-protein and thus swithces off adenylyl cyclase as in C.

Why have such elaborate controls for receptors? The answer probably lies in the deleterious effect chronic stimulation would have on cells. An activating mutation in a G-protein that stimulates adenylate cyclase (*gsp*, see Chapter 4) causes pituitary hyperplasia. As discussed in Chapter 3, the binding domain of receptors can be thought of as inhibitory in the absence of ligand. The conformational change elicited by ligand binding relieves this block. Indeed deletion of the extracellular domain of the EGF receptor partly underlies the tumorigenic potential of the v-*erb*B oncogene (see below).

Another desensitizing mechanism is that of auto-repression. Unlike the positive effect of c-*jun* on its own transcription (see above), c-*fos* inhibits transcription from its own promoter. Indeed the regulation of c-*fos* levels in cells is a model of efficiency. Within 10 minutes protein levels can increase by over 100-fold; by 30 minutes c-*fos* is barely detectable. Interestingly, the retroviral oncogenic form of *fos*, v-*fos*, does not *trans*-repress, which perhaps contributes to its transforming potential.

3. Concerted targeting

The reader might be excused for thinking that cellular regulatory processes are just too complex to contemplate. However, solace may be found in the limited number of ultimate targets for this apparatus. While there are many extracellular stimuli which operate through an array of specific receptors, many of the pathways that are ultimately affected are similar. Thus a diverse assortment of agonists such as bombesin, PDGF, insulin, phorbol esters, activated v-*src*, all increase the phosphorylation of one particular protein in cells termed p80. Why have such multiplicity in signals if the final result in the cell is the same? Of course, it is not the same. Each hormone or growth factor activates its own cocktail of effectors and it is the particular blend of pathway activation that delineates the specificity of reaction to a stimulus. Thus the responsiveness of a tissue to PDGF, for example, will depend on which cells have which type of PDGF receptor (α or β) and at what level. Of those cells, their particular response may depend on the expression of a particular isotype of one of the transducing proteins (which form of protein kinase C?). The immediate history of the cell with respect to other mitogens will affect the scale of effect. Ultimately the response may depend on which genes are physically accessible to the activated transcription factors with respect to the nuclear scaffold reflecting the cells' lineage and state of development.

3.1 Common end points

Certain cellular processes are affected by multiple stimuli implying a fundamental importance for these events (and justifying their considerable study). Such processes include ionic fluxes and the phosphorylation of certain key proteins.

Many mitogens cause intracellular alkalinization, which has been postulated to act as a signal. The pH change is caused, in part, by activation of a Na^+/H^+ antiporter in the plasma membrane, although the precise means by which this is achieved is still unclear. It is worth noting that the alkalization observed in the laboratory following activation of the Na^+/H^+

antiporter may not actually occur *in vivo* where the presence of bicarbonate buffering leads to compensatory ionic movements. Nevertheless, the experimentally observed activation of the antiporter still represents a common end point for which at least two distinct intracellular pathways have been defined (PKC-dependent and PKC-independent).

There are three proteins in particular that become phosphorylated in response to diverse mitogens. As mentioned above, a protein migrating on gels with a molecular mass of 80,000 (p80) is a prime substrate for PKC. As yet, little is known of the function of this protein despite the recent isolation of molecular clones. Activation of PKC is elicited by agents that stimulate the production of diacylglycerols (see Chapter 5) and phosphorylation of p80 has proven valuable as a cellular indicator for pathways acting through PKC. The two other proteins, ribosomal protein S6 and a 42,000 mass protein (p42), are phosphorylated in response to PKC-independent as well as dependent pathways. The latter protein is phosphorylated on tyrosine and threonine residues in response to growth factors and also in response to phorbol esters which directly activate PKC (a protein-serine kinase, see Chapter 5). Indeed, treatments that reduce cellular PKC levels decrease both threonine and tyrosine phosphorylation of p42. There must therefore be an intermediary tyrosine kinase or phosphotyrosyl phosphatase whose activity is modulated by PKC.

Ribosomal protein S6 is multiply phosphorylated at several serine residues near its C-terminus by the cyclic AMP-dependent protein kinase and at least two kinases "specific" for S6. These latter two protein kinases are inactive in resting cells and are switched on by being phosphorylated by S6 kinase kinases. There is accumulating evidence that the tyrosine/threonine phosphorylated p42 is identical to one of these S6 kinase kinases, thus explaining the coordination between increased p42 phosphorylation and ribosomal protein S6 phosphorylation. Activation of the p42 protein occurs only if both threonine and tyrosine residues are phosphorylated, thus the p42/S6 kinase kinase may act to integrate signals, transmitting a signal only if it senses two separate stimuli.

Another common response of cells is the rapid and transient expression of the immediate-early genes which include c-*fos* and c-*myc* (see Chapter 7 and above). c-*fos* is exquisitely sensitive to a wide range of perturbations including heat shock, stress, physical damage and even centrifugal force. It is something of a conundrum that this gene can be ubiquitously expressed in stimulated cells, yet is also capable of causing cellular transformation when in a deregulated state.

While phosphorylation of S6 has been reported to affect the translational capacity and selectivity of ribosomes (thus allowing mitogens to influence protein synthesis) and p42 appears to be an S6 kinase kinase, the functions of other common response proteins remain elusive. Although c-*fos* is clearly involved in transcriptional regulation, why should its expression be such a "knee-jerk" response to signals that are clearly not mitogenic? The promiscuous nature of c-*fos* induction highlights the role of gene regulation in many cellular responses, not just in cell division. Thus, excitation of nerves induces many of the same immediate-early genes induced by mitogens; these responses are multipotential. For example, under certain circumstances they may contribute to a proliferative response; under others they may be required in the replenishment of secreted proteins; it is the cellular context (i.e. the lineage history) that defines the outcome.

Elucidation of the molecular mechanisms by which these common, early responses are elicited will aid in our understanding of the way in which cells respond to growth factors and may provide an important target for drug intervention in neoplasia. Much effort has concentrated on the analysis of the mechanisms by which c-*fos* transcription is activated and repressed, as a paradigm for nuclear signal transduction. Regions in the c-*fos* promoter confering sensitivity to cyclic AMP, serum, phorbol esters and PDGF have been delineated, and the factors that bind to these regions are being characterized (see Chapter 7).

Despite the many questions that remain regarding these common target proteins, they provide invaluable markers for the activation of transducing pathways. The fact that they are targeted by such diverse effectors implies a fundamental role. Just as oncogenes have "showcased" proteins with key growth regulatory properties, these "common denominators" of cellular stimuli may well lead us to the basic elements of cellular coordination and control.

3.2 Multisite phosphorylation

Although the first phosphorylation event to be discovered involved the addition and removal of a single phosphate group on glycogen phosphorylase, many subsequent systems have been found to be rather more complex. Glycogen synthase, for example, the rate-limiting enzyme of glycogen deposition that opposes glycogen phosphorylase, is phosphorylated on nine serine residues *in vivo*. At least five different protein kinases act on the protein in skeletal muscle (figure 4A). What is the need for such complexity? The effect of each phosphorylation site on the catalytic activity of the enzyme is distinct: some are more inhibitory than others, while two appear to be silent (no effect can be detected *in vitro*). A clue to the apparent functional redundancy is gleaned upon examination of the effect of hormones on the individual phosphorylation sites (figure 4B). Adrenaline, for example, increases the overall phosphate content, but not equally throughout the sites. Indeed just two regions account for most of the increase. Since adrenaline elevates intracellular cyclic AMP levels and thus activates cyclic AMP-dependent protein kinase (see above), and this protein kinase phosphorylates glycogen synthase *in vitro*, it was believed to be directly responsible for the effect of adrenaline. This does not appear to be the case, however, for the sites that are affected by adrenaline are not exclusively the sites targeted *in vitro* by cyclic AMP-dependent protein kinase. The signalling must thus involve at least two activated protein kinases or inhibition of a protein phosphatase (see below).

Multisite phosphorylation acts as an integrator of signals. If several pathways impinge on the phosphorylation state of a protein of central importance, the activity of that protein will reflect the combined status of the impinging signals. There is not necessarily an on or off state, but a gradation. This is possibly the function of the "common end point" targets discussed above since ribosomal protein S6, and p42, for example are multiply phosphorylated. These proteins may thus integrate or "summarize" the information bombarding a cell and transform it into a scale of effect.

A.

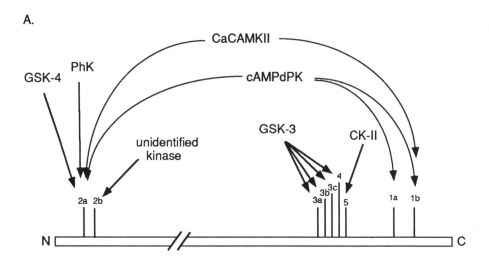

B.

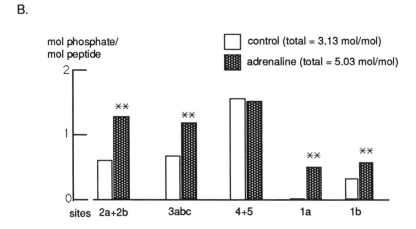

Fig 4. Phosphorylation sites of skeletal muscle glycogen synthase. A, Sites phosphorylated on glycogen synthase *in vitro* by cyclic AMP-dependent protein kinase (cAMPdPK), phosphorylase kinase (PhK), glycogen synthase kinase-4 (GSK-4), calmodulin-dependent protein kinase II (CaCAMKII), glycogen synthase kinase-3 (GSK-3) and casein kinase-II (CK-II). B, The histogram shows the changes in phosphorylation state of the various sites in glycogen synthase induced by adrenaline.

3.3 Concerted control - how to avoid shooting your foot

As mentioned in the regulation of glycogen synthase, increasing the phosphate content of a protein can be achieved by either activating a kinase or inactivating a phosphatase (or most efficiently by doing both). The reversibility of protein phosphorylation is a key feature of its utility as a regulatory mechanism. Indeed the half life of phosphate on proteins in cells can be very short, much less than the half life of the proteins they modify. Thus cells have developed a family of protein phosphatases to counteract the battery of protein kinases (see Chapter 6).

Since the kinase and phosphatases catalyse opposing reactions, they must be coordinately regulated to prevent futile hydrolysis of ATP. One mechanism that achieves cooperation between these enzymes centres on a phosphatase inhibitor (figure 5). This protein, termed inhibitor-1 (I_1), is only active, i.e. inhibitory, when phosphorylated on a threonine residue by the cyclic AMP-dependent protein kinase. The phospho-I_1 specifically inhibits the broad specificity AMD_c phosphatase (phosphatase1). Thus, agents that elevate intracellular levels of cyclic AMP not only increase phosphorylation of proteins that are direct substrates of cyclic AMP-dependent protein kinase, but through inhibition of the AMD_c phosphatase, also increase the phosphate content of proteins directly not targeted by this kinase.

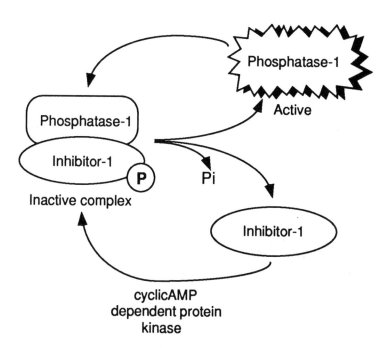

Fig 5. Inhibition of the AMD_c phosphatase (phosphatase-1) by cyclic AMP-dependent phosphorylation of an inhibitor subunit. Activation of cyclic AMP-dependent protein kinase leads to inhibition of a protein phosphatase and hence an increase in the phosphorylation of proteins.

These types of mechanisms can make elucidation of the molecular pathways of signal transduction very difficult. Evidence that a protein kinase phosphorylates a protein at the same residue *in vitro* that is phosphorylated *in vivo* in response to an agonist does not necessarily imply that the protein kinase is regulated by that agonist. Regulation may be at the level of a kinase with a similar specificity, a phosphatase or even the substrate.

One of the best examples of concerted targeting occurs in the regulation of cdc2. This gene product is the focus for multiple pathways that audit the cells' preparedness for division. As mentioned in Chapters 5 and 6, cdc2 is itself a protein-serine kinase which associates with several other proteins, such as the cyclins, in regulatory complexes. In yeast and probably also in mammals, cdc2 is negatively regulated by at least two other protein-serine kinases and a protein-tyrosine kinase. One might envisage progression through the cell cycle allowing sequential processing of cdc2, endowing the protein with the facets required for triggering entry into mitosis, but only when everything is ready. Thus the DNA must have been completely replicated, the cell must be large enough to divide, the microtubules must be organized, etc. Prior to completion all these processes must exert some sort of negative signal preventing premature commitment to division. The likely target of these completion signals is cdc2. Once all is set, cdc2 is activated by dephosphorylation (see Chapter 6) and rapidly triggers a series of events such as breakdown of nuclear lamins leading to chromosome condensation and segregation. cdc2 is then inactivated and ready to start sensing progress into the next cell cycle. The cdc2 protein kinase thus acts as a linchpin holding the mitotic process at bay until the vital preparations for division have been accomplished.

4. Cross-talk in cellular signalling

This chapter has already emphasized the importance of coordination in cellular regulation. Communication between pathways occurs not only at common end points as discussed above, but also at various levels within the signalling pathways leading to augmentation or suppression of signals.

4.1 Negative effects

Refractivity to a signal can be meeted out not only by homologous desensitization, but also in *trans*, by down-modulating the effects of other distinct agonists, an effect termed heterologous desensitization. For example, addition of PDGF reduces the ability of cells in culture to bind EGF. This transmodulation process occurs through post translational modification of the receptors which reduces their affinity for EGF. In some cell types, such as neutrophils, elevation of cyclic AMP inhibits polyphosphoinositide turnover. Conversely, phorbol esters acting through PKC inhibit β-adrenergic receptor coupling.

The desensitization of cells to agonists that use similar intracellular pathways results largely from the compensatory mechanisms that come into force after initial stimulation. The transitory nature of signals is a consequence of their rapid removal and destruction. Thus cyclic AMP is hydrolysed to AMP by phosphodiesterase activities; diacylglycerols are phosphorylated by diacylglycerol kinase; inositol trisphosphate is phosphorylated by an IP_3 kinase or dephosphorylated by the IP_3-phosphatase; calcium is taken up by organelles - these metabolic signals have their termination routes well organized. Indeed, when these mechanisms are overridden, such as by the poorly metabolized but potent phorbol esters and calcium ionophores, the entire growth control process may be compromised.

4.2 Positive effects

The interaction of different signalling systems is not always antagonistic. In many circumstances, agonists synergize with and even activate other pathways. In the cerebral cortex, activation of protein kinase C causes stimulation of the cyclic AMP pathway. In studies on the mitogenic response of Swiss 3T3 cells, the effect of combinations of growth factors and hormones has demonstrated classes of agents that can stimulate DNA synthesis only in the presence of other factors. Thus prostaglandin E_1 alone is without effect, whereas it is mitogenic in the presence of insulin, which likewise is impotent by itself.

One of the cell types present in the perinatal optic nerve is the O-2A progenitor cell. Under different culture conditions this cell type can differentiate into either oligodendrocytes (O) or type-2 astrocytes (2A). When optic nerve glial cells are dissected from a neonate rat and cultured in chemically defined medium, the O-2A progenitor cells stop dividing and rapidly differentiate into oligodendrocytes. Addition of PDGF to these cultures allows the progenitor cells to continue to divide and delays their differentiation until a point dependent on the age of the cells, at which time the cells synchronously form oligodendrocytes. This is believed to mimic the normal development of these cells in the rat. O-2A progenitors cultured in the presence of fibroblast growth factor (FGF) are able to divide but this growth factor is not able to delay the differentiation process (unlike PDGF) and FGF-treated cells thus rapidly form oligodendrocytes. Combination of PDGF and FGF causes the cells to divide indefinitely without differentiation; a very different scenario to the individual effects of the growth factors. This difficulty in predicting the combinatorial effect of growth factors from their individual responses is both the hero and villain of cellular regulation. On the one hand it makes for diversity and versatility; on the other it raises questions concerning the physiological relevance of studying the effects of growth factors in isolation.

While many of the underlying mechanisms of synergism remain to be resolved, much is known of the interactions between pathways. Thus agonists that elevate intracellular calcium affect other second messengers including cyclic AMP and diacylglcerol production. At the transcriptional level analysis of gene promoters has revealed the presence of multiple binding sites for proteins that can act positively or negatively depending on the context of the DNA (see Chapter 7). For example, the glucocorticoid receptor can activate or repress gene transcription of genes harbouring binding sites for this protein. As well as positively acting elements termed enhancers, regions of DNA that suppress expression have been identified, termed silencers. Whether, when and at what level a particular gene is transcribed depends on the interactions of these positive and negative forces, and it is reasonable to assume that synergy between growth factors acts partly through such regulatory elements.

4.3 Permissive effects

A common mechanism for cellular control is interdependency; that is where one event is required for, or greatly potentiates, the occurrence of a subsequent event. This type of regulation generates an ordered, temporal series of interdependent reactions. Several processes involved in cell cycle progression operate in this step-wise fashion, such as the requirement for completion of DNA replication before mitosis, as described above. The advantage of such control is the coordination of events that could be catastrophic if allowed to proceed without regard to other processes within the cell.

A molecular example of a permissive effect is the dependency of one protein kinase upon another for the formation of its substrate. A protein-serine kinase termed glycogen synthase kinase-3 (GSK-3) phosphorylates several proteins including glycogen synthase, a second inhibitor of protein phosphatas-1 termed the modulator protein, a protein that tethers a population of protein phosphatase-1 molecules to glycogen particles (termed G-subunit), and the transcription factor c-*jun*. In the case of the two former proteins, phosphorylation by GSK-3 is absolutely dependent on prior phosphorylation of neighbouring residues by an additional protein-serine kinase: casein kinase-II (figure 6). Cyclic AMP-dependent protein kinase appears to provide the "priming" phosphate for the G-subunit whereas no prior phosphorylation is necessary for c-*jun*. The sites targeted on glycogen synthase by GSK-3 are the most sensitive to hormonal changes (see figure 4). Insulin, for example, causes a specific decrease in their phosphate content accounting for most of the insulin-dependent activation of glycogen synthase. It is thus feasible that regulation of these sites is mediated through a pathway that does not directly affect GSK-3, rather the phosphorylation of the neighbouring site. If this were to occur, the level of phosphate in the casein kinase-II site should mirror that in the GSK-3 sites. However, this appears not to be the case since the phosphate content of the casein kinase-II site does not change upon hormonal challenge unlike that of the GSK-3 sites. In the case of glycogen synthase it would appear that this interdependency does not reflect acute changes but rather longer term alterations in the regulation of the enzyme: the same may not be true for the other examples of this phenomenon.

5. Cooperation in the action of oncogenes

As illustrated above, cells have developed many ways of utilizing signals to their fullest extent, to maintain control, to coordinate multiple messages and also to generate specificity of response. Since these are the mechanisms that are subverted in cancer, where do the faults occur and how many ways are there to cause cancer? The answer to this last question may be a large number, but it is not infinite. Cancer is a multistage process. One indication of this is the rapid increase in incidence of a number of cancers with age, suggesting the requirement for an accumulation of changes in a cell before it becomes cancerous. Evidence to support the idea that multiple events are needed to confound the intricacies of cellular regulation and thus cause neoplasia is discussed in the following section.

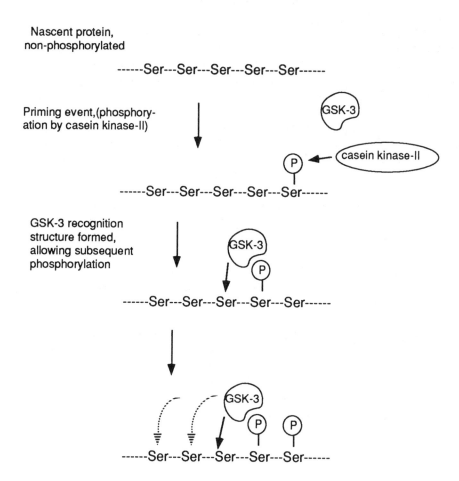

Fig 6. Cooperation between protein kinases. The diagram illustrates the cooperation between casein kinase II and GSK-3.

5.1 The two oncogene hypothesis

Although retroviral oncogenes cause tumours in the animals from which they are isolated, not all of the cells of the organism develop disease. This is not entirely due to the limitations of infectivity imposed by the particular virus. When new-born chicks are inoculated with Rous sarcoma virus, for example, tumours develop at the site of injection even though virus can be found in cells in other tissues. If the animal is inoculated at one site with virus and mock injected at a second site with saline, tumours develop at both sites. It thus appears that the trauma caused by injection contributes to the tumorigenic potential of the virus.

One of the properties of oncogenes is their ability to cause tumours in immuno-deficient mice. This test involves introduction of the oncogene, usually by DNA transfer, into cultured cells followed by implantation of those cells into the mouse. Tumours only form in mice injected with oncogene-expressing cells. However, transfer of such genes into primary cells, that is, cells that have not been previously cultured *in vitro*, does not support tumour development. This effect is explained by the phenomenon of immortalization. When cells are isolated from a tissue, a portion of the cell types undergo a number of rounds of cell division. After a certain time, the fraction of cells capable of division decreases until a point termed "crisis" when virtually all of the cells stop multiplying. These cells do not die but are senescent and refractory to mitogenic stimuli. The number of rounds of division that cells undergo before "crisis" is referred to as their Hayflick number and is species dependent: for murine cells it is about 30 and for human cells about 50. During crisis, a very small proportion of cells acquire the ability to divide *ad infinitum*. These cells rapidly overgrow the non-dividing senescent cells and are said to be "immortalized". The molecular basis for this immortality is poorly defined but is characterised by gross changes in chromosome number and appearance, which presumably reflect genetic changes altering the growth potential of the cells. Many commonly used cell lines have been generated in this way.

Expression of certain oncogenes can immortalize cells without requiring transition through crisis. The immortality of these cells is dependent on the presence of the oncogene. If it is inactivated the cells become senescent (indicating that they have in some way "counted" the intervening divisions and exceeded their Hayflick limit). Immortalizing oncogenes include SV40 T antigen, polyoma T antigen, adenovirus E1A, human papilloma virus E7, and *myc*. Interestingly, all of these proteins, with the exception of *myc*, associate with and functionally inactivate the retinoblastoma gene product, p105-RB, implicating this protein as a key determinant in the negative regulation of cell division (acting as an anti-oncogene, see Chapter 7).

A key observation pertaining to the multi-hit nature of cancer was the effect of introducing an immortalizing oncogene into a primary cell together with a non-immortalizing oncogene (figure 7). Such cells appear to be fully transformed. Thus *myc* alone immortalizes primary cells without transforming them. In contrast, introduction of an activated *ras* oncogene causes primary cells to develop a transformed phenotype (loss of contact inhibition, anchorage-independent growth, etc.) but without prolonging their lifespan. Co-expression of both *myc* plus *ras* causes the primary cells to adopt all of the characteristics of transformed cells, including the ability to form tumours in mice. Thus at least two independent kinds of change are needed to generate a neoplastic cell. However, studies in which transgenic mice carrying either *myc* or *ras* are interbred shows that *in vivo,* although there is an increased incidence of tumour formation, additional factors are also involved, indicating that at least three changes must occur to effectively bypass normal cellular controls.

Most acutely transforming retroviruses contain only one oncogene. However, the avian erythroblastosis virus has transduced two cellular genes termed v-*erb*A and v-*erb*B. The former is derived from the chicken thyroid hormone receptor but no longer binds the hormone and the latter is a truncated form of the chicken receptor for EGF. Deletion of the *erb*A gene still allows transformation by the virus *in vitro* via the v-*erb*B gene, but reduces

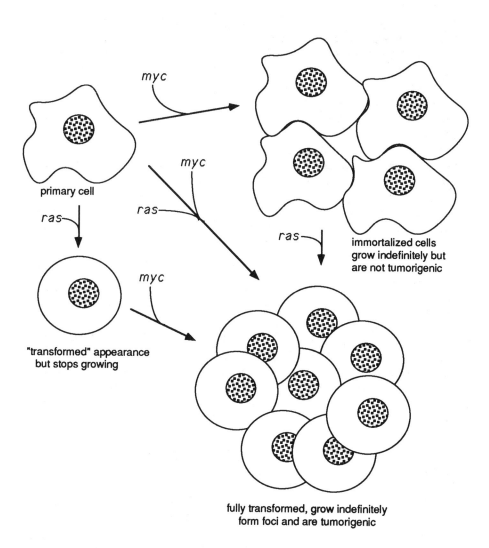

myc

primary cell

ras

myc

ras

myc

ras

immortalized cells
grow indefinitely but
are not tumorigenic

"transformed" appearance
but stops growing

fully transformed, grow indefinitely
form foci and are tumorigenic

Fig 7. Cooperation between oncogenes. The figure indicates the partial effects of the introduction of either *ras* or *myc* alone into a primary cell. The combination of the two however produces a fully transformed, immortal, tumorigenic cell.

its efficacy *in vivo*. *In vivo* the virus transforms erythroblasts which would normally differentiate into non-dividing erythrocytes; expression of v-*erb*A blocks residual differentiation of erbB transformed erythroblasts and thus maintains the population of dividing cells. This virus therefore has exploited the effect of cooperativity by transducing two genes with complementary functions.

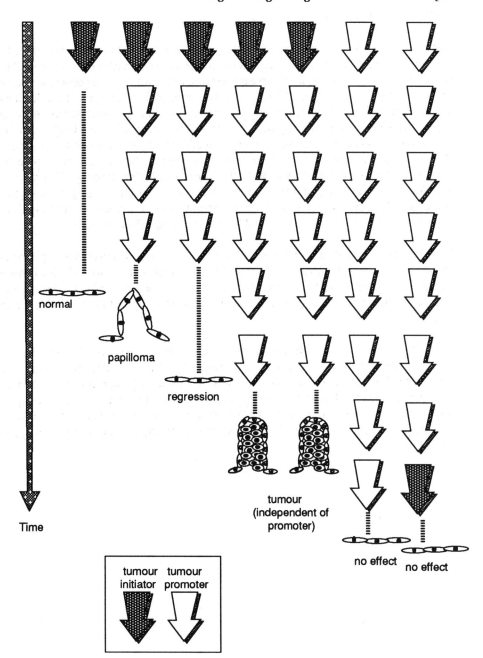

Time

normal

papilloma

regression

tumour
(independent of
promoter)

no effect   no effect

tumour   tumour
initiator   promoter

Fig 8. Initiation and tumour promotion in mouse skin. Initiation is accomplished by a single application of a subthreshold dose of a carcinogen. This step is irreversible but is not sufficient to induce altered cellular growth properties. Promotion is achieved by repetitive administration of a promoter as illustrated. Initially papillomas are formed which will regress if promotion is stopped. On continued promotion malignant tumours will form. The order of initiation and promotion is crucial. Thus either promotion alone or promotion followed by initiation is ineffective with respect to tumour formation.

5.2 Tumour promotion

The requirement for multiple changes in the generation of a carcinoma is also implied by the synergistic effects of certain tumorigenic compounds on mouse skin (figure 8). Some mutagens, termed initiators, do not give rise to tumours upon a single exposure but do cause genetic damage. This is revealed by the induction of tumours after repeated application of a second class of compound termed tumour promoters. These promoter agents only cause tumours if the skin has previously been treated with a mutagen. Repeated exposure of the skin to promoters followed by application of an initiator is also without effect, indicating that a particular sequence of events is required. Malignant tumours do not form immediately after treatment with promoting agents. Rather, small papillomas develop which regress if exposure to the tumour promoters is stopped. However, in time, a proportion of these papillomas become independent of the promoting agent and become cancerous. Thus several independent changes are needed to progress to full blown neoplasia.

The central involvement of signal transduction processes in the action of tumour promoters should be noted: protein kinase C (Chapter 5) and protein-serine/threonine phosphatases (Chapter 6) are potently affected by different classes of tumour promoter (phorbol esters and okadaic acid respectively), underscoring the importance of the tight regulation of these proteins.

The requirement for multiple events in the evolution of neoplasia is presumably a consequence of the intricate control mechanisms within cells designed to prevent run-away growth. As was stated at the beginning of this chapter, the vast majority of the cells in our body are not actively dividing. Those that retain the capacity to multiply must navigate a series of controls in order to do so. These inhibitory or fail-safe systems are our primary defence against cancer, and justify the massive investment cells make in evolving their regulatory systems.

9

Perspectives

1. Signalling networks in the control of cellular proliferation

The chapters in this book have covered various principles and examples of signal transduction particularly those involved in responses to mitogens. In the chart at the end of this book some of the signalling molecules that have been discussed are placed within current models of signal transduction pathways. The mitogenic action of some extracellular signals (e.g. adrenaline, angiotensin II) involves seven transmembrane receptors coupled to second messenger producing enzymes; as a consequence second messenger-dependent serine/threonine protein kinases become activated. Specific substrates for these kinases (i.e. cAMP-dependent protein kinase, protein kinase C, Ca^{2+}-dependent protein kinases) associated with mitogenic responses, have yet to be identified. In contrast to the above, polypeptide growth factors (such as EGF, PDGF and insulin) operating through receptors with tyrosine kinase activity appear to feed into multiple signal transduction pathways at various levels. The recent progress in defining substrates for these kinases has come from studies of signal transduction pathways themselves, leading to the identification of the type I PI-kinase, PI-PLCγ and GAP as substrates. At least in the case of PI-PLCγ this suggests that these receptor tyrosine kinases can circumvent a requirement for G-protein coupled activation of PI-PLC activity through direct interaction with this effector. In addition polypeptide growth factor signalling is likely to involve cascades of protein kinases since two of the potential *in vivo* substrates are serine/threonine protein kinases (MAP-2 kinase and *c-raf*). This type of direct interaction with intracellular kinases, in effect by-passes a requirement for second messenger generation. Other alternative strategies in signal transduction have been discussed in the action of specific extracellular agents; the extent to which these are involved in mitogenic signalling remains to be determined. Similarly other changes characteristic of mitogenic responses (e.g. ribosmal protein S6 phosphorylation and activation of Na^+/H^+ exchange) have yet to be placed precisely at common end points for these pathways.

A primary end point in mitogenic signalling involves stimulation of transcription and at least 100 genes are involved in this response. Several transcription factors (including CREB, c-jun and SRF) are thought to be important in this regulatory step and indeed some of the induced genes are themselves transcription factors (for example *c-fos* and *c-myc*) which may well be responsible for secondary changes in gene expression. While there is evidence for the "direct" coupling of CREB to the cAMP pathway it is not clear how other pathways feed in to regulate nuclear events. By contrast the action of steriod hormones is through direct association with intracellular receptors that are themselves hormone-dependent transcription factors.

It is evident that even within the boundaries of what has been discussed there are substantial areas of ignorance; beyond this it is likely that there are also as yet undiscovered pathways. Within the context of what is known, the major areas to be defined include the detailed elucidation of targets for the receptor tyrosine kinases (and phosphatases), substrates for the various serine/threonine protein kinases that operate at a lower level within the signal transduction hierarchies and an understanding of the links between cytoplasmic events and the nucleus. It is to be expected that in defining all these parameters, common elements will be revealed that rationalise the complex interplay between pathways. These interactions (as discussed in Chapter 8) form the basis of the normal physiological response of cells *in vivo* where individual hormones and mitogens are not exposed to cells one at a time.

An area not addressed here but nevertheless one which is likely to grow in importance concerns the signal transducing events triggered by agents that have a negative effect on cell proliferation. There are a number of factors and situations that have been described that appear to suppress cell division. Perhaps a clear cut experimental phenomenon would be 'contact inhibition' whereby primary cells growing in culture arrest when the dish is confluent (i.e. cells are in close contact). The nature of the signals that in effect desensitize the cells to mitogens have yet to be detailed, although one can speculate that one clue may have already been unearthed in the studies on the phosphotyrosine phosphatases (Chapter 6). The extracellular domain of one such phosphatase is related to cell adhesion molecules; it is possible that activation of this protein phosphatase through cell adhesion (which would be expected to antagonize effects of the growth factor receptor tyrosine kinases) may contribute to contact inhibition.

Notwithstanding the above, it is anticipated that in the not too distant future the signalling pathways that control cellular proliferation will be elucidated and mapped out in the same way that many regulatory pathways impinging on intermediary metabolism have been unravelled. The comparison of mitogenic signalling with the control of intermediary metabolism is an interesting one. It is now a widely accepted idea that the hormonal control of intermediary metabolism (i.e. extracellular influences) operates in conjunction with endogenous controls such as allosteric activation of enzymes by substrates. These endogenous controls serve to allow the cell to survive in the context of extracellular demands. In effect a similar level of control (albeit with a somewhat different purpose) operates with respect to cell division. Thus, for example, the cell has to monitor its own DNA synthesis in order that it does not enter mitosis too soon. Aspects of the complex regulatory systems operating to achieve this end are discussed in Capters 5 and 6. This plane of regulation and indeed much of the detailed biochemistry of the endogenously controlled processes on which extracellular signals operate remains a second general area of comparative ignorance. While elucidation of these "endogenous" controls and biochemical events is clearly of general importance, it remains to be seen if any of these elements have a causal part to play in human cancer (it may be anticipated that activating or inactivating mutations in these "endogenous" controlling genes will be simply lethal).

2. A unified view of signalling and neoplasia

The foregoing chapters have addressed in a stepwise fashion elements within presently known control hierarchies and put into context what is known of various situations where certain of these elements operate inappropriately, giving rise to uncontrolled growth. In reviewing the generalized scheme, outlined at the beginning of this book, it is evident that protooncogenes operate within control hierarchies at various steps. This is summarised in figure 1.

The nature of the alterations that lead to inappropriate action of these protooncogenes are diverse. The activation of a protooncogene may occur in the absence of a change in the structural gene; the inappropriate expression of a growth factor by a cell that is capable of responding to that factor would be an example of this (Chapter 2). The means by which this expression is itself achieved may vary. This can occur through mutations in the promoter region for a growth factor gene (or perhaps in the relevant transcription factors) or as evidenced by the mouse mammary tumour virus (MMTV) through integration of transcriptional enhancers within a growth factor gene (int-2; see Chapter 2).

A more subtle example is afforded by the ability of the viral protein middle T to complex with and activate the c-src protein (Chapter 5). In this instance the activation step is post-translational (although a genetic event is required for transformation since stable production of the middle T protein is needed). A similar scenario is afforded by a number of transforming proteins that have been found to bind to (and probably repress the function of) the retinoblastoma gene product (p105rb). p105rb is an example of a suppressor gene product whose positive action plays an important role in the negative control of cell proliferation; sequestration or mutation of this gene has dire consequences (see Chapter 7).

A much larger group of protooncogene activations is generated through minor to gross mutations in structural genes. The most simple of these is a single nucleotide change altering for example the glycine at position 12 in the ras proteins (Chapter 4). More drastic alterations also occur as evidenced by the gross truncations seen in the virally acquired oncogene v-erbB (see Chapter 3).

It should be clear from the detailed discussion in Chapter 8 that the operation of a signal transduction pathway does not occur in isolation. Thus all pathways are interconnected by positive and negative cross-talk and by certain common end points. It is therefore not surprising that combinations of events are required in vitro to obtain fully transformed cells. This in part reflects the complex nature of neoplasia in vivo where again multiple events are involved. An illustration of the cooperation between oncogenes is provided by the "natural" occurrence of the avain erythroblastosis virus, one isolate of which contains two genes of cellular origin, namely v-erbA (derived from the thyroid hormone receptor) and v-erbB (derived from the EGF receptor). These genes cooperate by blocking differentiation (v-erbA) and by stimulating proliferation (v-erbB).

One might ask whether it is coincidental or adventitious that various oncogenes (in their broadest possible conception) responsible for transforming cells, have to date been

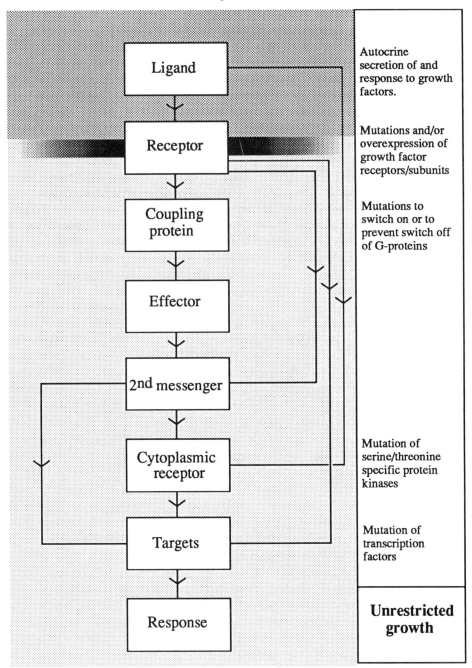

Fig 1. Generalized signal transduction and aberrancies. The general scheme for signal transduction outlined in Chapter 1 is reproduced with appropriate annotation indicating the levels within this hierarchy at which abnormal events leads to deregulated growth.

identified with elements involved in signal transduction. It would however be appropriate to ask first what we mean by transformation which for the purposes of discussion can be equated with cancer. It is clear that cancer is a disease of differentiated multicellular organisms. That is not to say that regulatory elements may not become altered in unicellular organisms (in the laboratory a variety of such mutants can be generated in, for example, yeast) but that the consequences of such alterations are limited. It is the differentiated nature of susceptible organisms that is critical. Such differentiation has evolved to create a series of specialised cell types that can cooperate in exploiting an environment and adapting to new ones. This cooperation requires communication to govern issues such as embryogenesis or morphogenesis and acute issues such as wound repair or energy supply. Cancer is, in effect, where one cell leaves the fold and starts on an existential crusade. This is of course not a positive decision but a consequence of a flaw in the normal process of communication that serves to integrate that cell in the organism as a whole. In this context it is perhaps not surprising that many genes that have been implicated to date in carcinogenesis do indeed fall into the category of regulatory proteins involved at one level or another in this communication process.

In the context of human cancer it is evident that the metastasis of cancerous cells is a major element in poor prognosis. While it is beyond the scope of this book to address the mechanisms of metastasis, it is pertinent to ask what types of genes may be involved in the ability of transformed cells to invade surrounding tissue, penetrate the endothelial wall, escape immune surveillance, and so disperse through the body. At one level it can be supposed that the loss of effective communication in terms of cell-cell contact is an important feature. This can therefore be accommodated in the general scheme of aberrant reception/signal transduction that has been the emphasis of this book. Beyond this however it may well transpire that quite distinct types of genes are involved in allowing cells to metastasize and it remains to be seen what the nature of these genes are. It is likely that any such gene alterations occur in combination with alterations in genes providing growth advantage and as such provide a rationale to the multiple events that are responsible for carcinogenesis.

So, aside from significant conceptual advances in understanding the "how" of cancer, where is all this getting us? The practical consequences of our present understanding will arise out of the tools obtained in reaching that understanding. One can ask specifically whether genes that have been defined experimentally as oncogenes (or protooncogenes) are in any way abnormally expressed in human cancer. It transpires that a number of such defined genes are indeed mutated or overexpressed. The extent to which such changes are causal remains in many cases to be determined, however these analyses are of prognostic value. Thus the overexpression of the *c-neu* gene (a growth factor receptor tyrosine kinase) in human breast cancer is associated with a poor prognosis and would indicate the need for a more aggressive chemotherapy regimen. Where there is evidence of causation in the inappropriate expression of a particular gene, it would be of therapeutic value to target that gene product pharmacologically. Whether such a selective approach will produce viable drugs for the future remains an open issue. However, based on our current knowledge of the signal transducing proteins involved in growth control and in particular of the multiple gene families that show selective tissue expression, it remains a fervent hope that suitable, tissue selective pharmacological agents will be developed.

Bibliography

Selected reading

Growth factors
- Sporn, M.B. & Roberts, A.B. *Peptide growth factors and their receptors volumes I and II.* Springer-Verlag. 1990
- Metcalf, D. The molecular control of cell division, differentiation commitment and maturation in haemopoietic cells. *Nature* 1989 **339**, 27.

Receptors
- Langer, J.A. & Pestka, S. Interferon receptors. *Immunology Today* 1988 **9** 393.
- O'Dowd, B.F., Lefkowitz, R.J. & Caron, M.G. Structure of the adrenergic and related receptors. *Ann. Rev. Neuroscience* 1989 **12** 67.
- Rubin, G.M. Development of the *Drosophila* retina. *Cell* 1989 **57** 519.
- Thomas, M. The leukocyte common antigen family. *Ann. Rev. Immunology* 1989 **7** 339.
- Yau, K.W. & Baylor, D.A. Cyclic GMP-activated conductance of retinal photoreceptor cells. *Ann. Rev. Neuroscience* 1989 **12** 289.
- Ullrich, A. & Schlessinger, J. Signal transduction by receptors with tyrosine kinase activity. *Cell* 1990 **61** 203.

Second messengers
- Sullivan, K., Tyler Miller, R., Masters, S.B., Beiderman, B., Heideman, W. & Bourne, H.R. Identification of receptor contact site involved in receptor- G protein coupling. *Nature* 1987 **330** 24.
- Casey, P.J., & Gilman, A.G. G-protein involvement in receptor-effector coupling. *J. Biol. Chem.* 1988 **263** 2577.
- Kobilka, B.K., Koblika, T.S., Daniel, K., Regan, J.W., Caron, M.G. & Lefkowitz, R.J. Chimeric α_2-, β_2- adrenergic receptors: delineation of domains involved in effector coupling and lingand binding specificity. *Science* 1988 **240** 1310.
- Tyler Miller, R., Masters, S.B., Sullivan, K.A., Beiderman, B. & Bourne, H.R. A mutation that prevents GTP-dependent activation of the α chain of G_s. *Nature* 1988 **334** 712.
- Krupinski, J., Coussen, F., Bakalyar, H.A., Tang, W-J., Feinstein, P.G., Orth, K., Slaughter, C., Reed, R.R. & Gilman, A.G. Adenylyl cyclase amino acid sequence: possible channel- or transporter-like structure. *Science* 1989 **244** 1558
- Rhee, S.G., Suh, P-G., Ryu, S-H. & Lee, S.Y. Studies of inositol phospholipid-specific phospholipase C. *Science* 1989 **244** 546.

Protein kinases
- Edelman, A.M., Blumenthal, D.K. & Kerbs, E.G. Protein serine/threonine kinases. *Ann. Rev. Biochem.* 1987 **56** 567.
- Nishizuka, Y. The molecular heterogeneity of protein kinase C and its implications for cellular regulation. *Nature.* 1988 **334** 661.
- Cooper, J.A. The *src*-family of protein-tyrosine kinases. *Peptides and Protein Phosphorylation.* CRC Press Inc. 1990.
- Hanks, S.K. & Quinn, A.M. Protein kinase catalytic domain sequence database: uses in identification of conserved features of primary structure and classification of novel family members. Methods in Enzymology "Protein Phosphorylation". 1991.

Protein phosphatases
- Ballou, L.M. & Fischer, E.H. Phosphoprotein phosphatases. *The Enzymes* Academic Press. New York 1986
- Waelkens, E., Agostinis, P., Goris, J. & Merlevede, W. The polycation-stimulated protein phosphatases: regulation and specificity. *Adv. Enz. Regul.* 1987 **26** 241-270.
- Cohen, P. The structure and regulation of protein phosphatases. *Ann. Rev. Biochem.* 1989 **58** 453.
- Cyert, M.S. & Thorner, J. Putting it on and taking it off: phosphoprotein phosphatase involvement in cell cycle regualtion. *Cell* 1989 **57** 891.
- Cohen, P. & Cohen, P.T.W. Protein phosphatases come of age. *J. Biol. Chem.* 1989 **264** 21435.
- Tonks, N.K., Charbonneau, H. Diltz, C.D., Kumar, S., Cicirelli, M.F., Krebs, E.G., Walsh, D.A. & Fischer, E.H. Protein tyrosine phosphatases: structure, properties and role in signal transduction. *Adv. Prot. Phosphatases.* 1989 **5** 149.
- Tonks, N.K. & Charbonneau, H. Protein tyrosine dephosphorylation and signal transduction. *TIBS* 1989 **14** 497.
- Cohen, P., Holmes, C.F.B. & Tsukitani, Y. Okadaic acid: a new probe for the study of cellular regulation. *TIBS* 1990 **15** 98.
- Steele, R.E. Protein-tyrosine phosphorylation: a glimmer of light in the darkness. *TIBS* 1990 **15** 124-.
- Prives, C. The replication functions of SV40 T antigen are regulated by phosphorylation. *Cell* 1990 **61** 735.

Transcriptional control
- Maniatis, T., Goodbourne, S. & Fischer, J.A. Regulation of inducible and tissue-specific gene expression. *Science* 1987 **236** 1237.
- Brave, R., Zerial, M. Toschi, L. Schurmann, M, Muller, R, Hirai, S.L., Yaniv, M. Almendral, J.M. & Ryseck, R-P. Identification of growth factor-inducible genes in mouse fibroblasts. *CSHSQB* 1988 **53** 901.
- Evans, R.M. The steriod and thryoid hormone receptor superfamily. *Science* 1988 **240** 889.
- Beato, M. Gene regulation by steriod hormones. *Cell* 1989 **56** 335.
- Green, M.R. When products of oncogenes and anti-oncogenes meet. *Cell* 1989 **56** 1.
- Johnson, P.F. & McKnight, S.L. Eukaryotic transcriptional regulatory proteins. *Ann. Rev. Biochem.* 1989 **58** 799.

- Wright, C.V.E., Cho, K.W.Y., Oliver, G. & De Robertis, E.M. Vertebrate homeodomain proteins: families of region-specific transcription factors. *TIBS* 1989 **14** 52.
- Berk, A.J. & Schmidt, M.C. How do transcription factors work? *Genes and Development* 1990 **4** 151.

Further reading

Cell cycle

- Lewin, B. Driving the cell cycle: M phase kinase, its partners, sexual preferences and substrates. *Cell* 1990 **61** 743.
- Murray, A.W. & Kirshner, M.W. Dominoes and clocks: the union of two views of the cell cycle. *Science* 1989 **246** 614.
- Nurse, P. Universal control mechanism regulating onset of M-probe. *Nature* 1990 **344** 503.

Oncogene cooperation

- Land, H., Parada, L.F. & Weinberg, R.A. Cellular oncogenes and multistep carcinogenesis. *Science* 1983 **222** 771.
- Thompson, T.C., Southgate, J., Kitchener, G. & Land, H. Multistage carcinognesis induced by ras and myc oncogenes in a reconstituted organ. *Cell* 1989 **56** 917.

Index

Contributors

Dr Paul Stroobant, Ludwig Institute for Cancer Research, Courtauld Building, 91 Riding House Street, London W1P 8BT.

Dr Julian Downward, Imperial Cancer Research Fund, 44 Lincoln's Inn Fields, London, WC2A 3PX.

Dr Josef Goris, Afdeling Biochemie, Dept. Human Biologie, Campus Gasthinsberg, Heresrtaat, 3000 Leuven, Belgium.

Dr Brian Hemmings, Friedrich Miescher Institute, Postfach 273 CH-4002 Basel, Switzerland.

Dr Silvia Stabel, Max-Delbruck-Labor in Der MPG, Carl-von-Linné Weg 10, D-5000 Köln 30, Federal Republic of Germany.

Dr Nicholas Jones, Imperial Cancer Research Fund, 44 Lincoln's Inn Fields, London, WC2A 3PX.

Dr James Woodgett, Ludwig Institute for Cancer Research, Courtauld Building, 91 Riding House Street, London W1P 8BT.

Dr Michael Waterfield, Ludwig Institute for Cancer Research, Courtauld Building, 91 Riding House Street, London W1P 8BT.

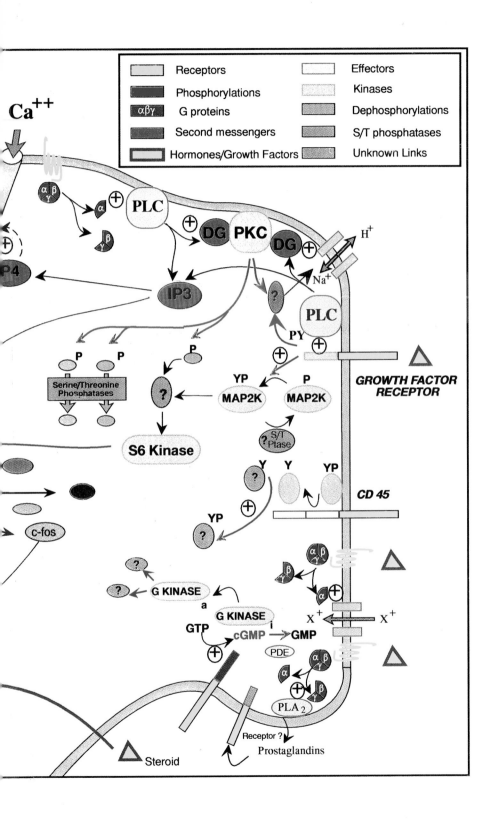